THE GESTAPO MANUAL

THE GESTAPO AND SS MANUAL

TRANSLATED BY
CARL HAMMER

PALADIN PRESS
BOULDER, COLORADO

The Gestapo and SS Manual
Translated by Carl Hammer

ISBN 0-87364-875-7
Printed in the United States of America

Published by Paladin Press, a division of
Paladin Enterprises, Inc., P.O. Box 1307,
Boulder, Colorado 80306, USA.
(303) 443-7250

Direct inquiries and/or orders to the above address.

TABLE OF CONTENTS

PREFACE

THE GESTAPO (SECRET STATE POLICE) OF HITLER'S GERMANY was hardly known to the public as a police force whose mission was to serve and protect. Nobody, however, has yet argued that the Gestapo was an incompetent police organization.

The methods of the Gestapo are not only of historical interest. They are also of some real relevance today to both those who are interested in police tactics and operations and those who fear the danger posed to a free society by excessive abuse of police privileges. *The Gestapo and SS Manual* provides a rare glimpse into the operating procedures of the Gestapo, a subject until now virtually undocumented.

These methods were used and perfected before and during the World War II. The rules and procedures detailed in this handbook formed the standard operating procedure of the Gestapo and the SS police. This handbook, or the experiences of being exposed to methods described herein, also formed the foundation for the postwar classic, *Total Resistance*, written by the Swiss Major H. von Dach Bern in 1958 (also available from Paladin Press). It is interesting to compare the two books, as the *The Gestapo and SS Manual* was written for the use of the secret police, while the other was written for the use of the resistance fighter. The two books complement each other in many ways.

Although the Gestapo was most known for its secret police role, not all operations were of a secret nature. Raids were common, and crowd and riot control—although entrusted to regular police, SS, and army units under Gestapo control—was always important. The present handbook is, therefore, devoted to operations involving raids on political opposition meetings and local resistance headquarters and to the dispersal of political demonstrations, protesting mobs, or rioters. There is also a chapter on paramilitary operations against partisans and resistance fighters.

The first part of the handbook describes the fundamentals of police training. German police basic training in those days was not dissimilar to regular army training. The second part describes paramilitary operations in open terrain, that is, in the country as opposed to the cities. Particularly, partisan and resistance activities often took place away from the big cities. The book's third part details operating procedures for urban police operations in times of peace and war. This part details operating procedures that have never before been made public.

The first impression when reading this handbook is its clear structure and order. With customary German thoroughness, the text begins with rules on how to read maps and then continues with operations in open terrain. Operations in the city environment—for which the Gestapo and the other German police organizations were most notorious—are covered last. It is apparent that the author or authors of this handbook saw it as their duty to begin with the easy parts and then to move on to more complex situations.

When reading this handbook, one also notices some of the changes that have taken place since the 1930s, when *The Gestapo and SS Manual* was written (no exact publication date is given, and it appears that the present copy is one of several editions; besides, all illustrations were printed separately and then pasted into the handbook itself). No longer are European roads usually in poor condition, and neither partisans nor police rely on horses anymore. Yet many of the methods and techniques described in this handbook remain as valid today as when they were first written.

But *The Gestapo and SS Manual* was not only for the Gestapo and the Nazi German police; it was also for the Waffen-SS.

The regular police forces often worked together with the Waffen-SS, and indeed many police officers were transferred to this organization. The paramilitary methods described in this handbook thus came to be used not only against partisans and resistance fighters, but also against the regular-army opponents of the Waffen-SS. Contemporary sources describe the huge losses generally suffered by the markedly more aggressive Waffen-SS and SS police units when compared to those of the regular Wehrmacht units. Wehrmacht General Erich Hoepner once even condemned the Waffen-SS' disregard for casualties as "a butcher's mentality." The tactics and fighting methods of the Waffen-SS and the Wehrmacht certainly differed in many aspects. This handbook, therefore, also provides some glimpses into the operational procedures of the Waffen-SS itself, and not just those of the Gestapo and regular police forces that made up the bulk of the internal security apparatus of Nazi Germany.

INTRODUCTION

•

A BRIEF HISTORY OF THE GESTAPO AND THE POLICE OF NAZI GERMANY

SAY THE NAME GESTAPO, AND YOU WILL MOST LIKELY THINK of an evil organization of secret police—a secret thought police, designed only to inspire dread, terror, and discipline based on fear. And these associations are correct. Such were indeed the main purposes of the organization known—in practice although not in fact—as the Gestapo.

Reinhard Heydrich, the ruthless head of the Reich Main Security Office, once said that the Gestapo was supposed to inspire a "mixture of fear and shivers." The cold, dispassionate bureaucrat Heinrich Müller, head of the Gestapo itself, would not have disagreed. Yet the Gestapo had its origin in a perfectly legal political police organization, a police department that in the early days—before it was made part of the Schutzstaffel (SS)—fought against the criminal excesses of the Nazi organizations (the brown-shirted SA—Sturmabteilung—of Ernst Röhm and the black-shirted SS of Heinrich Himmler) as often as it struggled against the terrorist activities of the Communists.

The history of the Gestapo is the history of a police department that gradually became subverted by a political party. Even in 1945, there were officers in the Gestapo who remembered the time when the Nazis were the bad guys and the democratic government was on the side of the angels. To them, this did not matter because the Nazi Party was in power for many years, and theirs was but to obey instructions from the government. Nobody protested when the government awarded the Gestapo increasingly wider responsibilities, legal powers, material resources, and privileges. The path from a democratic police force to an evil organization was not obvious and was not traveled in one day.

Heinrich Müller, SS-Gruppenführer and Lieutenant General of Police, Amtschef (Head) of the Gestapo.

A multitude of books have been written on the illegal and quasi-legal activities of the Gestapo and its part in the extermination of Jews and other minorities in Hitler's Germany. Let us here—for a change—content ourselves with sketching a brief history of the Gestapo, with the emphasis on the actual police activities of the organization rather than with its more murderous aspects. But let us not fool ourselves: the Gestapo had blood on its hands—and was proud to admit it.

In Nazi Germany, the security apparatus, of which the Gestapo was the chief part, was all encompassing. As many as 45,000 officials and employees of the Gestapo, divided among 21 principal posts and 36 branches within the Third Reich itself, as well as another 300 principal posts and 850 posts within the border police (Grenzpolizei) formed the core of the security apparatus. There were also the security organs provided by the SS and the uniformed police: 65,000 security police officers and

2.8 million uniformed police. The more than half-million inmates of the 20 concentration camps and 160 associated labor camps were guarded by 40,000 guards. Finally, the Sicherheitsdienst (SD), or Security Service, controlled no less than 100,000 informers, all of them eager to report the most intimate thoughts of the citizens. The informers were everywhere, in every factory, corporation, farm, university or institute of learning, and, of course, in every government organ. All reports were duly sent to the Berlin headquarters.

One must also not forget the almost one million soldiers of the Waffen-SS, of whom more than 300,000 were ethnic Germans from southeastern Europe and another 200,000 were foreigners, with no particular loyalty to the German people. The Waffen-SS was not under the authority of the military forces and could be used as an armed wing of the Nazi Party in case of disputes with the military.

The security apparatus of the Third Reich was certainly strong but did not develop overnight or according to a careful plan. Indeed, the history and development of the Gestapo show a sketchy and almost haphazard progress. This was no coincidence. Anybody who believes that the Third Reich was built on long-term planning, careful order, and a strong hierarchy is fooling himself. Instead of authority and order, the internal organization of the Third Reich was characterized by an almost absolute lack of hierarchy and structure. This was deliberate; the Führer (leader), Adolf Hitler, feared competition, and indeed any strong counterforce that could prevent the exercise of what he called the dynamic will of the leader.

Hitler, accordingly, encouraged a continual power struggle among his chief followers. Any action could be justified, whether it had been ordered or not, as long as one could say that it was merely the execution of the will of the Führer. And Hitler was deliberately vague in expressing his dynamic will. The quasi-hierarchy of the Nazi party, the military, and the security apparatus counted as nothing compared to the will of the Führer.

For this reason, these three forces—the party, the military, and the security apparatus—more often than not worked at cross-purposes. The secret police, too, often worked against other Nazi organizations. But this was not all: any section also had ruthless competition among its own members. Not even the Gestapo was immune to these internal power struggles. Many of its sections opposed each other and engaged in fearsome struggles. As for the lesser components of the security apparatus, such as the Criminal Investigation Police (Kripo—Kriminalpolizei) and the uniformed police, they were increasingly dominated by the powerful Gestapo. The history of the Gestapo is also the history of how one part of the security apparatus managed to completely dominate all its fellow services and almost the entire state.

RUDOLF DIELS AND THE EARLY GESTAPO

The Gestapo began not as one, but as two separate police organizations. One—the older—was the political police of Munich, in the state of Bavaria. The other—the younger organization, but the one whose name lived on—was the political police of Berlin, in the state of Prussia.

By the early 1930s, Germany had experienced more than its fair share of political violence. After losing World War I the country was poor, with rampant unemployment. Socialist, Communist, and various Nationalist parties battled for power. Money and resources were scarce, but after the war weapons were plentiful. To get information on the many political extremists, who threatened law and order in often violent street battles, the weak authorities tried to ride out the storm by setting up political police sections within ordinary police departments. As the German states were more or less autonomous, each one organized its own police force.

In this climate of violence and poverty, Adolf Hitler (1889–1945) rose to power. Following a free election, he was appointed chancellor on 30 January 1933, and the Nazi Party seized power in Germany.

The new Führer did not wish to lose political power as easily as his opponents had before him. He appointed his close followers to all important posts. One of these was the much decorated combat pilot ace and swashbuckling war hero Hermann Goering (1893–1946).

Goering, although by birth a Bavarian, was made minister of the interior in Prussia, commander in chief of the Prussian police, and, by the way, commissioner for aviation. Within the Prussian police Goering's ascension to power caused far-reaching changes. First, he immediately created his private armed force, a "police force for special purposes," known as Landespolizeigruppe General Goering. Second, he fired no fewer than 1,457 officers and officials from the police as suspected enemies of the Nazi regime. Finally, Goering reorganized the political police section. Within the Prussian State Police (Stapo), Section I A of the Berlin police was responsible for political intelligence. Goering chose this section as a foundation for a new political police. The new political police force was established by Goering on 26 April 1933 and placed directly under the authority of the Ministry of the Interior. As a new name, he chose the appellation Geheimes Staatspolizeiamt (Secret State Police Department).

However, the new police force yet existed only on paper; it also needed flesh and blood. As new director of Section I A, Goering in June 1933 chose Rudolf Diels (1900–1957), a boisterous drinker and womanizer, who had shown his talent and ambition ever since he first entered the Prussian Ministry of the Interior in 1930. In 1932, Diels had been appointed senior government councillor in the political police section of the ministry. Because of his powerful sense of observation and shrewdness, he displayed great talents for this work.

It is unclear whether it was Goering or Diels who first came up with the idea of a political police with increased responsibilities and powers. The two men were certainly good friends. Some said that Diels was the confidant of Goering because of his skill in the stock market, with which he assisted the luxury-loving Goering. Diels was also one of the first police officials to join forces with the Nazis. In any case, Diels built Section I A of the Stapo into the new secret state police that Goering desired. Diels' section became a strong foundation for what was to become the Gestapo.

The abbreviation Gestapo was not yet in common use. The official abbreviation of the new political police force, invented by an unknown post office worker who designed its stamp, was Gestapa. In the usage of the German people, however, this friendly-sounding name soon changed into the word that all Germany was to fear: Gestapo.

The Gestapo, or Geheime Staatspolizei (Secret State Police), was an appropriate name. In German, *geheime* means secret but also private. The new political police was certainly secret. In due time, it also became private: a private police force serving one party—and indeed one individual.

Diels not only had to find men to staff the new secret police, he also had to locate offices large enough for the projected size of operations. He soon found suitable headquarters at 8 Prinz Albrechtstrasse in Berlin. Several large buildings, including the Museum of Folklore and a school, were evacuated and occupied by the new secret police.

Subordinate bureaus were also created in each district of Prussia and were placed directly under the Gestapa, the secret state police headquarters. This was the first major expansion of the activities of a Prussian secret police, the old political police having been limited to Berlin, Prussia's capital.

The Gestapa was the administrative center of the new political police. The actual political police—the core that became the Gestapo—formed Amt (department or office) III of the Gestapa, while the security police formed Amt IV. Amt I was responsible for organization and administration, whereas Amt II was responsible for judicial matters.

Before the Nazis' assumption of power, the political police of the various German states had not hesitated to fight the Nazis, together with other extremists. Nazi Party members had, by the end of 1932, been condemned to no less than a total of 14,000 years in prison and fined one and a half million marks. One of those who had contributed to these scores was Diels. Although the Nazi Party now was in charge, the political police of Diels at first did not hesitate to continue the struggle against Nazi violence. Diels especially concentrated his forces on the illegal activities of the brown-shirted SA of Ernst Röhm. In some way, Diels had learned that torture was frequently used in the SA headquarters. This was clearly illegal, and Diels appealed to Goering for permission to clean up the SA mess. Goering, anxious not to offend other Nazi leaders, hesitated but finally gave in to Diels and the Gestapo.

From then on, Gestapo officers armed with machine guns surrounded first one and then many others of the SA headquarters. The SA did indeed have unofficial concentration camps and prisons, where torture and maltreatment of political opponents constituted a frequent problem. The Gestapo cleaned out most of the establishments, releasing the inmates. Many SA members were arrested.

The SA, however, was not the only Nazi organization that kept illegal prisons. The SS of Heinrich

Himmler (1900–1945) also kept illegal concentration camps and prisons, including one large building in Berlin, the Columbiahaus on Papestrasse. The SS proved to be a more difficult problem for the Gestapo, as Himmler was rising fast in the Nazi leadership. Despite this, Diels and the Gestapo officers, not all of them Nazis, worked hard to dissolve the illegal SS prisons. But the SS was too strong. The Gestapo had little success, and, eventually (after Diels had left the organization), the secret police was instead joined to the SS. By then, the Columbiahaus became the private prison of the Gestapo, remaining outside the control of the Ministry of Justice.

Goering, meanwhile, had realized the power of having a properly organized political police, operating without any scruples, on his side. The trouble was, however, that somebody had to authorize the Gestapo to become suitably unscrupulous. Goering decided to see to this himself. In a number of speeches, he outlined what he saw as the proper conduct of a secret police force.

As early as 17 February 1933, Goering addressed the police officers of Prussia, instructing them to "not hesitate to shoot in case of need. Every agent must fully understand that inaction is an error more serious than a mistake committed in the execution of received orders." A police officer was to show no mercy, he declared. Goering also assumed responsibility for any excesses: "Every bullet that now exits the barrel of a police pistol is my bullet. If you use it for killing, I am the killer; I have ordered all this; I take it on my conscience. I assume full responsibility, and I am not afraid of it."

On 3 March in a public speech, Goering, as minister of the interior of Prussia and commander in chief of the Prussian police, continued: "My job is not to render justice; my sole goal is to destroy and exterminate, nothing else."

Two days later, on 5 March, Goering also became prime minister of Prussia.

On 24 July 1933, Goering finally proclaimed: "Anybody who in the future raises his hand against a representative of the National Socialist (Nazi) movement or the State should know that he will lose his life with little delay. It will even be sufficient to prove that he had the *intention* of committing this act, or when the act is committed, it did not lead to death but only to a wound."

This declaration was not the last of Goering's aggressive speeches, but it was the one which—to the Gestapo and eventually to all German police forces—publicly authorized unnecessary violence in the fighting of enemies of the Nazi regime. Other prominent Nazis soon repeated this instruction. One of the principal Nazi lawyers, Gerland, who promulgated instructions to the German judges, wrote that one "must again respect the term terror in the penal code."

Diels, although a Nazi, friend, and protégé of Goering, remained a police official of the old school. It was clear to anybody except the self-confident Diels that his days as head of the Gestapo were approaching an end. As the Gestapo became increasingly powerful, the Nazi leadership desired another type of director. Despite this, Diels continued to build the power and influence of his organization. Goering warned his friend: "I warn you, Diels, you want to sit on two chairs at the same time!" Diels replied, smiling: "The director of the Gestapo must sit on all chairs."

This advice, although said jokingly, was something that the future directors of the secret police would heed.

Diels did not remain head of the Gestapo for long. Caught in the power struggle between Goering and the SS leader, Himmler, Diels soon found himself on the receiving end of the police power. The first indication was a dramatic incident in which several members of the SS led by Herbert Packebusch stormed the home of Diels, locked his wife into the bedroom, and forced open the locked desks in Diels' private office. Packebusch was a friend of the powerful Kurt Daluege (1897–1946), a former Nazi thug and street fighter, who presently became commander of all Prussian police forces, as well as commander of SS in eastern Germany. Mrs. Diels, however, managed to call her husband by the bedroom telephone. Soon Diels, furious, returned home with a police troop. Packebusch, surprised, was still busy looting Diels' office. He tried to draw his service pistol, but the efficient policemen grabbed him before he managed to pull the trigger. The SS men were arrested, but because Daluege was closely connected to Goering, Packebusch had to be released to avoid trouble.

This was the beginning of the end for Diels. Two weeks later, at the end of September 1933, Goering gave in to the other Nazi leaders and ordered units of uniformed police and the SS to surround the home of Diels, who was forced to leave his post and fled to Karlsbad, Czechoslovakia.

Goering now had to find a successor to Diels. He nominated an old Nazi, Paul Hinkler. Unfortunately, Hinkler was a confirmed alcoholic, as well as a former murder suspect, with the reputation of an imbecile. He was clearly not the strong leader Goering desired as head of the secret police. Less than 30 days after Hinkler's nomination, Goering recalled Diels from Karlsbad to resume his duties as head of the Gestapo.

Goering wanted an efficient chief of the Gestapo, but he also wanted to rely on the secret police for his own purposes. On 30 November 1933 Goering, therefore, by decree placed the Gestapo under his personal direction instead of that of the Ministry of the Interior. This was, however, merely a delaying action. As will be seen, the Gestapo was by then slipping out of his hands.

REINHARD HEYDRICH, THE SD, AND THE BAYPOPO

The state of Bavaria, and especially its capital Munich, was a stronghold of the Nazis. One of the pillars of the Nazi power was the SS and its leader, Heinrich Himmler.

The SS had been established in early 1929 as the personal bodyguard of Adolf Hitler. From a small body of black-shirted men, the organization soon grew in size and power. Its leader, Heinrich Himmler, was eventually appointed Reichsführer-SS (Reich Leader of the SS).

Since the creation of the SS, every SS unit had included two or three men who were in charge of "security." Their task was to provide intelligence on the activities of any enemies, especially the Communists. In early 1931, Hitler ordered Himmler to establish a security service to protect the Nazi Führer. Himmler accordingly created a Referat (Section) Ic, so named after Section Ic of the old German general staff, responsible for intelligence on the enemy. The new security and intelligence service soon grew in size and importance. Ic sections were established in all local SS organizations. Each was ordered to report to the central Section Ic.

As Himmler reorganized the security service of the SS, he began the search for a security officer. After an introduction from a mutual friend, on 14 June 1931 Himmler met a former German navy lieutenant, Reinhard Heydrich (1904–42). Himmler asked the 27-year-old Heydrich to write, in 20 minutes, a first sketch of a future intelligence service. The talented Heydrich was equal to the task, and Himmler appointed him head of the new security service a few months later. Although Himmler retained the formal leadership of Section Ic, Heydrich soon took responsibility for all practical work.

Heydrich was tall, slim, and blond, with deep-set blue eyes. He was also a first-class athlete, a fencer, horseman, and pilot. With his military bearing, Heydrich was the perfect "Nordic Aryan type" of Nazi propaganda. He was also a talented violinist and a disciplined and merciless intelligence officer with experience from naval intelligence.

Forced to resign from the navy after an intimate relation with a shipyard director's daughter, Heydrich joined the Nazi Party in July 1931. He soon also joined the SS, where he was appointed Sturmführer on 10 August 1931. From 1 October 1931, Heydrich applied his experience from naval intelligence to his new service. He rose rapidly through the ranks. He was promoted to Hauptsturmführer (captain) on 1 December 1931, Sturmbannführer (major) on 25 December 1931, then formal chief of the SD (the new name for the security service) on 19 July 1932, and Standartenführer (colonel) on 29 July 1932. This was followed by the rank of Oberführer (brigadier) on 21 March 1933, and SS-Gruppenführer (lieutenant general) on 1 July 1934.

In April 1932 (that is, before the Nazi Party seized power) the government tried to suppress the SS and the SA. To overcome this threat, Section Ic of the SS briefly changed its name to the Presse- und Informationsdienst (PID, or Press and Information Service). The threat to suppress the SS did not last long, however, and Himmler and Heydrich were free to rename the organization with a more suitable, military-sounding name. The new name they chose was to become almost as notorious as the SS itself: the Sicherheitsdienst des Reichsführers SS, or the Security Service of the Reichsführer-SS, abbreviated as SD.

Heydrich developed the idea that whereas the former police had been satisfied with catching the enemies of the state in the act and had only acted in case of danger, Heydrich's security service was to identify and track down enemies even before they dared to think any dangerous thought and certainly before they dared to commit any act of opposition. The security service—and its extension, the police—was no longer to be a defensive intelligence organ of the state; it was supposed to lead the offensive against all enemies and indeed serve as the all-powerful educator of the people, to rid the nation of all undesirable ideas.

In practice, SD members were to occupy all key positions in the new police, which was then to be released from the control of the Ministry of the Interior. In the end, this is also what Heydrich achieved.

Heydrich and especially his protector, Himmler, had, however, a yet higher objective, which they never successfully realized: to combine the police, the SS (of which the SD was a part), and the entire civil service into a totalitarian SS state, patterned after the knights of the Teutonic Order of the Middle Ages. This was Himmler's ultimate reverie. But Himmler always remained a dreamer. The subordinate, but intellectually superior, Heydrich usually knew how to ensure that his superior made the right decision. When discussing plans with Himmler, Heydrich's every suggestion was a masterpiece of logic. Himmler, who seldom could follow the forceful arguments, usually gave in and let Heydrich have his way. Only once did Himmler get angry. He shouted: "You . . ., you and your logic. You always come here with your logic! Whatever I suggest you brush aside with your logic!" Heydrich retreated gracefully and, instead, had his way at the next meeting. Hitler, the Führer, often referred to Heydrich as "the man with a heart of ice."

After the Nazi seizure of power in Bavaria on 9 March 1933, Himmler was appointed police president of the Munich police. On 1 April, he also became commander of the political police of Bavaria, a new organization within the Bavarian Ministry of the Interior based on the old political police. The new organization became known by its abbreviation: BayPoPo.

Because Himmler was satisfied with Heydrich's work within the SD, he appointed Heydrich as his representative in the Bavarian police headquarters. Himmler and Heydrich then reorganized the BayPoPo as a new political police and enrolled many SD members. Heydrich then successfully separated the new organization from the administrative control of the police headquarters in Munich. As Himmler's assistant in securing control of the police, Heydrich brought all local political police groups under the control of the new BayPoPo and acquired the right to call in the uniformed police of Bavaria as an executive organ, whenever required.

The result was a thoroughly Nazi-dominated political police organization. In faraway Berlin, Diels had managed to retain at least a notion of equality before the law. This was not the case in Bavaria. The minister of the interior, a rabid Nazi, had already established concentration camps for political prisoners, and these camps were now subordinated to the BayPoPo. The Nazi rulers of Germany—"to protect the people and the state"—had already empowered the police, "as a preventive measure," to detain any citizen in a concentration camp. In Bavaria, this blank authorization was regularly used.

Bavaria was, however, only one of 17 German states. Among the Nazi leadership, the opinion on how to continue with the others was divided. Himmler, the Reichsführer-SS, wanted to destroy the police structure in the German states and instead create a national "state police" that stood above the local state structure.

Goering, however, strongly opposed Himmler, because Goering already controlled the Prussian political police, an independent organization in many ways similar to the BayPoPo.

Goering had one strong advantage: one of his close associates was Kurt Daluege, lieutenant general of police, who since May 1933 had been in charge of the police division in the Prussian Ministry of the Interior and commander of all Prussian police forces. Daluege was also the commander of SS units in eastern Germany and Himmler's main rival in the SS.

Himmler, however, was not alone in opposing Goering's dream of having a private police force. In September 1933, Himmler was made commander of all political police units outside Prussia. After the election on 12 November 1933, Dr. Wilhelm Frick in the Reich Ministry of the Interior decided to abolish the last independent powers of the German states. From then on, they were supposed to remain as administrative units only under a central administration in Berlin. This reduction of power naturally also concerned the police. From then on, no written procedure, promotion, or appointment (from the rank of major up) was allowed without the approval of the Reich Ministry of the Interior.

Goering was too fast for the centralists. On 30 November 1933, before Dr. Frick's proposition had become law, he promulgated his own law that raised the Gestapo of Prussia to an independent branch of the state administration. The Ministry of the Interior of Prussia no longer had any authority to relinquish the Prussian Gestapo to the Reich authorities.

Outside Prussia Dr. Frick's law soon took effect. Himmler assumed control over the police forces in each state. In November 1933 he was appointed director of the political police in Hamburg, the second city of the Reich; Lübeck; and Mecklenburg-Schwerin. In December he assumed control over Anhalt, Baden, Bremen, Hessen, Thüringen, and Württemberg; in January 1934 he added Braunschweig, Oldenburg, and Sachsen. By January, when Dr. Frick's new law was laid before the Reichstag (parliament), Himmler controlled the political police organizations in all German states except Prussia and Schaumburg-Lippe. Later in the spring only the police of Prussia remained.

Goering still resisted but eventually had to give in. By the end of March 1934 he surrendered most of the control over the Prussian police. Though formally under Goering, on 20 April 1934 Himmler became head of the Prussian police and assistant director and inspector (in other words, de facto head) of the Gestapa. The Gestapa, it will be remembered, formed the administrative and control organ of the Gestapo. Rudolf Diels, the first head of the Gestapo, had already been dismissed on 1 April 1934.

The commander of all Prussian police forces as well as commander of SS in east Germany, SS-Gruppenführer Kurt Daluege, the other associate of Goering, became commander in chief of the uniformed police of the entire Reich. He was also promoted to SS-Obergruppenführer (general). Daluege's friend Arthur Nebe (1894–1945?), a professional police officer and head of the Berlin criminal investigations section, as well as the author of an authoritative treatise on criminology, was appointed head of the Kriminalpolizei (Criminal Investigation Police—Kripo) of the Reich. Daluege also received command of the security police (so far part of the Gestapa, but distinct from the Gestapo) of the Reich and Prussia, and presumably regarded the political police as his next goal. This ambition was soon thwarted, however.

The SD chief, Reinhard Heydrich—as the assistant director of the assistant director Himmler—moved to Berlin and was immediately appointed de facto director of the Gestapa. He remained head of the SD and in control of the BayPoPo. This was the beginning of a new era for the Gestapo. The various political police organizations of Germany were from then on united, and the Gestapo assumed responsibility for operations throughout Germany.

Being in control of both the SD and the Gestapo, Heydrich had become a powerful man. The SD, too, had received increased powers. On 9 June 1934, the Führer's office had decreed that henceforth there was "no longer to exist any security service beside the SD of the Reichsführer-SS and not any intelligence service of the Party, nor any interior intelligence service for foreign political purposes." The SD assumed responsibility for all these functions.

The able organizer and administrator, the pedantic Himmler, was now in total control of the most powerful sections of the German police. For the Gestapo, this was sufficient, and the organization from then on steadily grew in importance. For Himmler, however, this was not enough. But he had the means to increase his power further. On 17 June 1936, Himmler was appointed supreme chief of all German police forces, including the political police, the Kripo, and the uniformed police, throughout the Third Reich, becoming the formal head of the Gestapo in addition to his position as Reichsführer-SS.

THE NEW GESTAPO AND HEINRICH MÜLLER

Reinhard Heydrich was ruthless and calculating, but he was also a professional. He did not care so much about the political views of his subordinates as about their loyalty and professional ability. Many such men were found in the Kripo in Munich. Having been transferred to the BayPoPo, many of these officers, including Heinrich "Heini" Müller (1901-?), were of dubious Nazi loyalty and only expected to be fired. After all, they had spent much of their time fighting the Nazis until the change of political power in 1933. Especially Müller, a short, stocky, but decorated World War I pilot, employed by the police ever since the end of 1919, had reasons to worry. He was a cold, bureaucratic fanatic, with a dry, expressionless face and a reputation of being the most virulent Communist-hater in the entire Bavarian Kripo. Although Müller had a thorough knowledge of police work, he was known to be indifferent to the Nazi ideology.

Müller and his colleagues looked down on Heydrich and the SD, which in their eyes was an upstart organization with few, if any, professional qualifications. But Heydrich needed professional expertise. Not only did he allow Müller and the others to retain their posts, he soon promoted them and invited them into the SD. Especially Müller was rapidly promoted, to SS-Standartenführer (colonel) in 1937, and Oberführer (brigadier) on 20 April 1939. Heydrich wanted to buy their loyalty and experience.

To some extent, Heydrich succeeded. Müller and his Bavarian colleagues did soon become thoroughly loyal, but to the Nazi regime, not to the SD. In later years, when Müller became the director of the Gestapo, he successfully resisted the SD's right to investigate the Gestapo's internal affairs, even though he himself remained an SD officer. (Despite this, Müller was further promoted to Brigadeführer [major general] on 14 December 1940, and Gruppenführer [lieutenant general] and chief of police on 9 November 1941.)

In Berlin, too, there was a group of professional police officers who joined the new secret police. One of them was Arthur Nebe, a ruthless and ambitious opportunist who delighted in his lack of a formal higher education. He had joined the Kripo after World War I and reached the rank of police commissioner in 1924. As had many police officers, he joined the Nazi Party and the SS in July 1931, attracted by the firm Nazi stance in favor of the Kripo, the promises of more funds and better technical equipment, protection of the police against a critical free press, and tougher criminal laws. Nebe also transferred to the Gestapa, where for a while he served as director of the executive branch, Amt III, in what was to become the Gestapo.

The third major group of professionals who joined the new secret police consisted of professionals, administrators, and lawyers from all over Germany. Their self-appointed leader was Dr. Werner Best (born 1903), a doctor of law from Heidelberg University and a judge in Hessen. A Nazi from an intellectual point of view, Dr. Best was made police commissioner in Hessen in March 1933. He had to resign from this post in the autumn; instead, he was appointed SD leader in the south and southwest. Dr. Best also was appointed director of the (reorganized) Gestapa Hauptamt I (main office, administration and judicial matters), and in the summer of 1933, also Hauptamt III (the police intelligence service). In this position, he used his legal skills to justify the totalitarian practices of the Nazi state and to give legitimacy to the secret police and the concentration camps. By 1935 Dr. Best was already an SS-Standartenführer (colonel).

Heinrich Müller and the Bavarians brought in by Heydrich formed Hauptamt II, the actual Gestapo. Bavarian specialists led the most important directorates of this branch. Heinrich Müller was appointed director of Section II A, the section in charge of Marxism and persecution of the illegal Socialist and Communist parties. In this position, he was also second in command of the Gestapo, after Heydrich. Franz Josef Huber was put in charge of the section investigating reactionaries, the judicial opposition, and the church. The rabid Nazi Josef Meisinger headed the section for expulsions from the Nazi party and racial disgrace, with the task of exposing homosexuals, apostates, Jews and non-Jews who engaged in intimate relations (especially after the Nuremberg racial laws of September 1935), and party members opposing Hitler. Willi Flechner headed the economic section, in charge of investigating legal associations and financial groups, including the German Labour Front (Deutsche Arbeitsfront, DAF). Reinhold Heller, a survivor from the time of Diels, was put in charge of the section for illegal trade unions. Dr. Karl Hasselbacher headed the section for Freemasonry and religious sects. There was also an increasingly important hunt for saboteurs—and in due time any error or mistake at work could be construed as sabotage.

The Gestapo had received a new organization and, to some extent, new areas of operation, but a legal framework for the new secret police was still missing. This minor omission was soon rectified. On 2 May 1935, it was decided that the secret police of Prussia were no longer subject to judiciary control. This principle was enshrined in law on 10 February 1936, when Goering as president of the Reichstag and prime minister of Prussia promulgated a "fundamental law" of the Gestapo. This law stipulated that the Gestapo had the duty to search for all forces hostile to the state, throughout Germany. The law also stipulated that "the orders and internal affairs of the Gestapo are not subject to examination in any administrative tribunal." From then on, there was no appeal against an order from the Gestapo.

The first article of the fundamental law stated: "The Gestapo has the mission to inquire into all intentions that put the State at risk, and to fight against them, to analyze and exploit the results of such inquiries, to inform the government, to keep the government informed of important facts, and to supply it with prompting to act upon these." Note that only an *intention*—and not any particular action or plan—to put the state at risk was sufficient to involve the Gestapo. The organization thereby received unlimited powers.

In the introduction to the decree of 17 June 1936, in which Himmler was appointed the supreme chief of all German police forces, now unified, the Nazi concept of a police force was spelled out, stating that "having become National Socialist (Nazi), the police no longer have the mission to protect an order established by a parliamentary and constitutional regime. The police exist (1) to execute the will of a single leader, and (2) to preserve the German people from all attempts of destruction by internal and external enemies. To reach this goal, the police must be all powerful."

Himmler reorganized the police into two main branches, the ORPO (Ordnungspolizei), or Public Order Police, and the SIPO (Sicherheitspolizei), or Security Police.

The ORPO consisted of several branches and organizations, including the Schupo (Schutzpolizei) or Protection Police, the urban uniformed police; the Gendarmerie or Rural Police; the Verwaltungspolizei or Administrative Police; the River and Canal Police; the Coastal Police; the Fire Service; and the Passive Defense and its Technical Auxiliary Police. Himmler, in his first order, signed on 25 June 1936, consigned the entire ORPO to his old rival, SS-Obergruppenführer (general) and General of Police Kurt Daluege. He already was chief of the old Security Police, by then part of the central office of the SD and responsible for protecting the lives of Adolf Hitler and the other Nazi leaders.

The ORPO enjoyed high visibility and retained some of its former high prestige, but the SIPO was the senior—and more powerful—service. The SIPO consisted of the Gestapo and the Kripo. All investigative services were turned over to this plainclothes force. In the same first order, Himmler consigned the SIPO to Reinhard Heydrich, who also was appointed second in command after Himmler. Heydrich in his turn turned over the leadership of the Gestapo to Heinrich Müller, who had been its de facto chief since 1935. The Kripo was, as before, entrusted to the ambitious police chief Arthur Nebe. Heydrich himself retained the title of director of the SIPO and the SD.

The high-level changes of the SIPO meant little for the lower-level officers. There the Gestapo and the Kripo still maintained separate offices.

The ORPO not only lost in influence, it also lost manpower. Of the original 150,000 men, a third joined the military, while the rest remained in the ORPO. Most were apparently happy with the Nazi regime, as the formerly low pay and dangerous duties were replaced by Nazi "law and order," significantly higher pay, better possibilities for promotion, and—some said—more attractive uniforms.

A system of card indexes and files was developed already by the Gestapa. It was expanded and improved upon. In both the Berlin headquarters and the local branch offices, all known enemies of the state were registered in the so-called A index. This card index consisted of three groups of cards. Group A 1, all cards of which were recognized by a red mark on the left side of each card, was reserved for enemies of the state, who being already in the preliminary stages of a secret mobilization were to be apprehended. Group A 2, with blue marks, consisted of people who were to be arrested upon the proclamation of a mobilization. Group A 3, meanwhile, consisted of citizens who were not enemies of the regime but were, as Heydrich put it, "in times of hard tests [to] be regarded as politically so dangerous that their apprehension and special surveillance must be considered."

Each card also had a mark on the right side. A dark red mark signified a Communist, a light red mark a Marxist, a brown mark a terrorist, and a violet mark a complaining troublemaker. The official in charge of the card index had to go through it twice a year, on 1 April and 1 October, to ensure that each classification remained valid.

The filing system was the most modern of its kind in Germany, and perhaps in the world. Half a million cards—the most important and interesting individuals from the point of view of the Gestapo—were placed in a huge, horizontal circular card index machine, an early version of a punched-card reader. The machine had an electric engine, which could speedily find any card. The complex mechanism was handled by only a single operator.

The files included extensive dossiers not only on enemies of the party and the state, but also on Heydrich's (and Himmler's) personal rivals and colleagues. Such files were more than once used to set members of the Nazi leadership at each other's throats. No fact was too small to be included in the files and the card index.

The task of updating each card naturally demanded a steadily increasing number of employees and resources. The Gestapo had begun with only 35 men from Section I A of the Berlin police. In the beginning of 1935, the Gestapo already counted 607 officials and other employees. From 1933 to 1937, the budget rose from one million to 40 million Reichsmarks.

The card index and file system was not the only invention of the Gestapo. Around 1936, Heydrich created an even more amusing kind of intelligence-collecting establishment. The Gestapo and SD officer Walter Schellenberg later related that the notorious womanizer Heydrich had the habit of frequenting the high-class brothels and Berlin houses of assignation. In one of these, he realized that the women working there knew much of what was going on, as their customers often gave them remarkable confidences, not only on the personal level—which might be important enough—but even on the most exalted level of politics. No doubt, the customers believed that the women only listened out of politeness and soon forgot the conversations.

Heydrich decided to take advantage of this. Through an intermediary, he rented a luxurious hotel and established it as a top-class house of assignation, the Salon Kitty. The technicians of the Gestapo and the SD installed hidden microphones in every room and in the secluded areas of the bar. They also planted listening devices in the walls and in selected furnishings. Arthur Nebe, the professional police officer, had once worked in the vice squad. He was therefore charged with recruiting suitable young ladies to work in the establishment. Not only charm and beauty were required, but also intelligence, culture, linguistic abilities, and—naturally—a high degree of patriotism. According to Schellenberg, some young ladies of the upper classes even volunteered for this kind of work, for patriotic reasons.

Salon Kitty soon became the favorite haunt of especially foreign diplomats, whose German "friends" never failed to recommend the establishment. Much valuable intelligence was collected. Heydrich himself frequently visited his creation. Naturally, he saw that the microphones were turned off whenever he made a personal appearance.

In 1937, the Gestapo branch sections throughout Germany grew into centers directly connected with the local government. The local branches were generally known as Stapo posts, but they served directly under the Gestapo. The Gestapo also received more responsibilities in the field formerly held by military counterintelligence. Eventually, the German borders also fell under the control of the Gestapo. The border posts had until then been under the authority of the local branch of the Kripo; now each was put under the authority of the Gestapo. Each one became a Gestapo post for foreign intelligence. Later the Gestapo also created its own Grenzpolizei. Not only did this service guard the borders; it also hunted the enemies of the state there. It therefore fell under the control of the Hauptamt III (the police intelligence service) of the Gestapo.

While the Gestapo drew an iron curtain around the borders of Germany, a detailed system of issuing warrants for arrest developed. No fugitive was to be allowed to leave Germany, and within the Reich, he was to be arrested speedily. The local police organizations retained their ordinary warrants, but a new type of warrant was introduced from the Stapo. Here too different colors were used to characterize different types of wanted criminals. Brick-red cards were used for escaped prisoners. White cards with red borders were used for those who were to be expelled. The Gestapo divided the warrants for arrest into eight groups:

- Warrant A, indicating arrest
- Warrant B, meaning arrest if the individual did not have a fixed address
- Warrant C, indicating the need to get a residence permit
- Warrant D, meaning expulsion
- Warrant E, being used for missing persons
- Warrant F, signifying the search for missing documents
- Warrant G, indicating secret control
- Warrant V, being used to arrest habitual criminals

Officially, the executive powers of the Gestapo encompassed three kinds of sanctions, which could be imposed without any need for a court order or trial: the right to warn, the right to arrest and impose protective custody, and the right to imprison suspects in concentration camps. Included also, naturally, were such executive powers as the right to interrogate and to confiscate evidence of various kinds. These legal sanctions even enabled the Gestapo to lawfully rearrest an individual on his way out of court, where he had just been acquitted of a previous charge, and then to put the individual in a concentration camp. It eventually became common to simply place any arrested suspect in a concentration camp until the formalities and paperwork of, for instance, expelling stateless people, had been sorted out.

This was sinister, as it was not unusual for concentration camp prisoners to simply disappear while in the camp. In such a case, there was, of course, no need for a trial or the presentation of evidence. Later—in preparation of and during the war—the Gestapo also at times used recently killed concentration camp prisoners as dummies in intelligence operations that, for some reason or another, demanded a freshly killed corpse. Such prisoners, still alive but earmarked for killing, were known as "canned food" in the Gestapo vocabulary.

Although the Gestapo's legal powers were vast, the organization also habitually resorted to illegal methods. Kidnappings, assassinations, and murders disguised as accidents or suicides were common means to fulfil the will of the Führer.

By the end of 1938, the Gestapo was fully developed. The organization had throughout the Reich no fewer than 57 regional offices of the Gestapo, divided into 21 Stapo *Leitstellen* (principal posts) and 36 Stapo *Stellen* (posts).

In addition, the Kripo (which since 1936, together with the Gestapo, formed the SIPO) controlled 66 regional offices, divided into 20 Kripo Leitstellen and 46 Kripo Stellen.

THE POWER OF THE GESTAPO

In Nazi Germany, the state was the Nazi Party. Party officials such as Heydrich had considerable powers. The result was a perplexing mixture of state and party organizations, orders, and instructions, made more confusing by the constant intrigues and personal antagonisms among the Nazi leaders. There was no unified leadership, with the single exception of Adolf Hitler, the Führer, and he constantly played the various officials off against each other, constantly changing orders and opinions. It was no wonder that Germany lost the war; the wonder was that it lasted as long as it did.

Since the beginning of 1933, Germany was divided into 32 *Gaue*, or administrative districts. Every *Gau* was divided into *Kreise*, or circles, every *Kreis* into *Ortsgruppen*, or groups of municipalities, every *Ortsgruppe* into *Zellen*, or cells, and every *Zelle* into *Blocks*. Each division was commanded by a leader, known respectively as *Gauleiter*, *Kreisleiter*, *Ortsgruppenleiter*, *Zellenleiter*, and *Blockleiter*.

The key to the Gestapo control system was the Blockleiter. Each Blockleiter, and there were vast numbers of them, controlled and was responsible for a block, that is, at most 60 households. He was the only official who was in direct and constant contact with all elements of the population. A Blockleiter had many responsibilities. He had to explain the laws to his subordinate citizens, whenever necessary. He also had to identify and report anybody who spread unsuitable rumors or committed crimes against the Reich. His responsibilities encompassed both the personal and professional lives of his charges. He was the one who knew what was going on in the Reich.

These activities led to a system in which the state expected each citizen to spy on his or her neighbors—and be spied on by them. Since 22 June 1933, when Goering signed an instruction to every official to report any behavior deemed contrary to the interest of the Reich, this was the rule. Not to inform was regarded as an act hostile to the state.

Each such report found its way to the Gestapo.

Every factory and every business company in the Reich also had a Nazi Party organization. Since early 1933, the monolithic labor organization known as the German Labor Front (Deutsche Arbeitsfront, DAF) grew into the largest, single mass organization in the Third Reich. It eventually controlled 25 million German workers. The DAF—led by the uncouth alcoholic, ardent Nazi, and close friend of Adolf Hitler, Robert Ley—assumed the role formerly played by the trade unions, which were now outlawed. According to an instruction from Goering dated 30 June 1933, the officials in this organization had the duty to inform the Gestapo of any worker whose political attitude seemed doubtful.

The organization of German farmers under the Reichsbauernführer (Reich Farmers' Leader) and pig breeder Walther Darré, an incompetent romantic, controlled the farmers as Robert Ley controlled the workers. Here, too, there was a duty to report any suspicion.

There were also other organizations of various kinds, both in the Reich and eventually in the occupied areas. These included youth organizations and cultural or vocational groups. The latter included associations of lawyers, medical doctors, students, and the Chamber of Writers and the professional association of professional writers. Only members of one of these had the right to have their works published. The leaders and officials of any organization were naturally required to report any unsuitable or merely interesting traits among their members.

Outside the borders of the Third Reich, information passed to the Gestapo from several sources. There was the Auslands Organisation (AO), or the Organization Abroad), of the Nazi Party, which comprised no fewer than 350 groups worldwide, and the Volksdeutsche Mittelstelle (VOMI), or German Racial Assistance Office, controlled by the SS, which was busy organizing the return to the Reich of anybody of German blood. The Gestapo found these organizations especially useful when operations were being planned against Austria and Czechoslovakia.

There was also the Auslands Politische Abteilung (APA, or the Bureau of Foreign Politics) of the Nazi Party. This bureau not only had the well-known duty to spread Nazi propaganda abroad, it also had a section of foreign press translators. Every day, this section provided a translated press review and extracts from 300 foreign newspapers. Much information of use to the Gestapo was also disseminated, especially information concerning political exiles. This included such information as notices of marriage, births, deaths, conferences, commercial enterprises, and much else. This section also kept records on the political orientation of foreign journalists.

In 1933, Hermann Goering also established the Research Institute Hermann Goering. Its main duties were to tap telephone switchboards and to control telegraph and radio transmissions. The activities of the institute were led by specialists from the navy, with assistance from a number of police officers, including Rudolf Diels, then still chief of the Gestapo. Attention was devoted primarily to international telephone, telegraph, and radio traffic. It was even possible at times to eavesdrop on communications between two foreign countries, as several countries used the German facilities for transit purposes.

Inside the Reich, specialists from the institute also tapped conversations between important individuals. The same happened to known foreigners and all those suspected of political deviation or under the surveillance of the police. Any line could be tapped and the conversation recorded. Even the telephone conversations of Adolf Hitler were routinely tapped.

Because the institute remained under the control of Goering, it gave him considerable power, and he refused to surrender his control over it. Nonetheless, the Gestapo and the SD could use the services of the institute, whenever necessary.

The Gestapo also installed its own wiretaps. Under the pretext of repairing or checking the telephone wires or electrical circuits, Gestapo agents installed listening devices in every suspect's home. Nobody was immune—even Hjalmar Schacht, then minister of finance, once found a hidden microphone in his home. He then discovered that the Gestapo had recruited his chambermaid to record his conversations. Goering, too, despite his personal eavesdropping institute, found himself wiretapped by the Gestapo and the SD.

THE NEW SD

The Nazi Party now had a firm control over Germany. Reinhard Heydrich, as director of the SIPO and the SD, was a man of importance, at least because of his SIPO connection. The SD, however, had done little if anything to secure the Nazi Party's control over Germany. The SD was still small, consisting of only a hundred employed SD officers and another hundred with honorary positions within the organization. Few people really knew what the SD did, and yet fewer realized that the SD did very little. The newly employed, 28-year-old SD agent Otto Ohlendorf (1908–1951), an intelligent, idealistic young lawyer and economist, who later rose to prominence in the organization, joined only because he thought the SD was an intelligence service. He later said: "The first disappointment was that in 1936 there was not yet any organ within the SD which worked with intelligence matters." Despite this surprise, the intelligent Ohlendorf did well. He was promoted to SS-Sturmbannführer (major) in 1938.

Ohlendorf belonged to the new generation of young intellectuals who now entered the SD. This group revolved around Walter Schellenberg (1910–1952), a bright and adventurous young man who had already worked for the SD while he studied law and political science at the University of Bonn in the early 1930s. After graduating, Schellenberg joined the Nazi Party and the SS in May 1933. He entered the SD in 1934. He soon also joined the Gestapo, where he—promoted to SS-Sturmbannführer (major)—was appointed head of the counterintelligence section.

The young SD intellectuals were exactly what Heydrich needed. Until then, he had been dependent on the Gestapo, which, although highly effective, retained the traditions of the old German and Prussian administration. Heydrich wanted to revolutionize the way in which the security police worked. The police officials and lawyers of the Gestapo, however, were quite content to maintain the old order of things, even though they now had new masters.

Heydrich wanted all his subordinates to be flexible, eager, able, and without any unnecessary concerns about legal procedures. Dr. Werner Best, the doctor of law from Heidelberg University and former judge, had instead demanded that only those with a legal background should be appointed top officials. Dr. Best, SD leader in the south and southwest and director of the Gestapa Hauptamt I and Hauptamt III, was not without influence. Heydrich, however, was of the opposite opinion, and so was the Führer. Hitler had once even said, "I will not give up until every German realizes that it is shameful to be a lawyer."

In the beginning of 1935 Heydrich therefore split the SD in two. This was not an administrative division—the SD remained one organ—but it was a clear split in the ideological role of the organization. The first half was SD as a party organ, to which all members of the security police were to be joined (thereby also joining the SS). The second half was SD as an intelligence organization, as the "mobile instrument, the sensory organ of the people's body, within all enemy groups and in all fields of life," as Schellenberg once explained it. It was therefore no longer mandatory, nor indeed certain, that a member of the SD be a member of the intelligence organization of the SD. On paper, however, there was no formal distinction between the two functions.

By this move, regular but old-fashioned members of the SD (such as Dr. Best) and members of the SD appointed only through their position (such as Müller and Huber) were retained as members of the SD only in its first role—that is, as a party organ. They were, however, excluded—and even resented and bitterly fought against—as members of SD in its role as an intelligence organization with an unlimited area of responsibility.

The SD had become the intelligence service of the German Reich. Its office, the Sicherheitshauptamt, was set up on Wilhelmstrasse in Berlin. Heydrich, the director, remained at Prinz Albrechtstrasse, where he could keep an eye on the Gestapo.

The headquarters of the SD was divided into three sections: Section I concentrated on organization, Section II managed the struggle against internal enemies, and Section III oversaw foreign operations. Any subsections were named after the main section's number, but with the addition of Arabic numerals.

The SD headquarters controlled seven regional SD commands, each of which controlled two or three subordinated centers. Each center in turn controlled additional branch offices, each responsible for a district or a major city. These branch offices formed the most important cells in the intelligence network of the SD in Germany.

The SD relied on a vast number of informers. The volunteer informers were known as *V-Männer (Vertrauens Männer*, or men of confidence). Only a few of these were members of the party or the SS. Even critics of the Nazis were at times enlisted as informers to provide intelligence on critical attitudes toward the Nazi regime. The informers were a heterogeneous group; some were minor employees and workers, but others were well-known judges, captains of industry, artists, and scientists. The SD had a special liking for academics, many of whom joined either as informers, employees, or honorary SD members. University professors who did not become informers instead found themselves under the surveillance of informers recruited among their students. During the lectures, the latter took notes that were then turned over to the SD for use in judging the political attitudes of the professors. Many women also served as informers, especially during the later years of World War II.

Every informer and agent had a special file at SD headquarters on Wilhelmstrasse. The file of each agent and informer was maintained by two controllers. One controller knew the agent's cover name, biography, loyalty, and details of his achievements and where he was to operate; the other knew the reasons for the agent's missions and graded his reliability, value, and worth according to a scale from 1 (excellent) to 5 (absolutely worthless). The agent's real name, however, was not disclosed to these controllers. This piece of data existed only in the central card index and in the card index kept at the local branch office in the area where the informer operated.

After the Nazis seized power, the SD grew rapidly. After the middle of 1934 the SD already employed 3,000 members. By 1937 the SD also had 50,000 informers. A few years later the SD controlled no fewer than 100,000 informers.

The only question was, what exactly was the SD to work with? The SD, unlike the Gestapo, did not have executive powers. And the Gestapo already did a more than efficient job in locating and eliminating enemies of the state.

The responsibilities of the SD overlapped those of the Gestapo. Agents of the two agencies frequently interfered with each other's work. As an example, Heinrich Müller's Section II A of the Gestapo (Marxism) hunted the same enemies as Section II 121 (Leftist Organizations) of the SD.

Himmler, therefore, soon found a new purpose for his SD. On 1 July 1937, Heydrich ordered that the Gestapo was to be responsible for the areas of Marxism, national traitors, and emigrants, while the SD was to concern itself with science, the life of the people, art, education, the party and the state, the constitution and administration, foreign countries, Freemasonry, and public associations. The SD was thereby transformed into a kind of spiritual police, an instrument for thought control. Real crimes were not as interesting as the lack of proper spirit and an eager Nazi attitude. A typical SD report from 1938 noted that a soldier on temporary leave, Fritz Schwanebeck, had "while singing the national anthem through a slack bearing showed a lack of interest." The implication was obvious: soldier Schwanebeck was not a good German citizen and Nazi.

The SD also worked hard—and successfully—to find out the identity of anybody who voted against the Nazi Party or merely left a blank vote in the elections. Each such disloyal individual soon had a new file at SD headquarters.

Many disputed areas of responsibility among the Gestapo and the SD remained, however. These included the church, religious sects and associations of religious and ideological nature, pacifists, Jews, reactionaries, farms, the press, and "other groups hostile to the state." It was, therefore, decided to split the area of responsibility, so that the SD worked on the general and fundamental questions only, while the Gestapo worked on individual cases in which executive powers were required.

The Gestapo therefore assumed all important powers, despite the formally higher position of the SD. There was, however, one field in which the ambitious SD man could excel without treading on the toes of the Gestapo: foreign intelligence.

Although the young SD intellectuals often had enjoyed fantasizing about foreign intelligence, any actual work in this field originally grew out of the coincidence that the SD also followed the enemies of the regime abroad. The idea of penetrating and infiltrating foreign countries as part of a war effort arose only later.

In the early operations, the young SD intellectuals did not show much professionalism. Most or maybe even all of them, from Heydrich on down, showed more fascination with methods they had read about in popular spy and detective novels than with the boring police procedures of the Gestapo or the rigid military methods employed by the military intelligence service (Abwehr). Heydrich especially admired the British intelligence service and even assumed the code name "C," which—he had read somewhere—was the mysterious code name of the head of the British intelligence service.

Most of the early SD foreign intelligence operatives acted with more enthusiasm than skill. One of the first was the former mechanic Alfred Naujocks (1911–1960). Although he had briefly been an engineering student in Kiel, Naujocks was a ruffian, brawler, and well-known amateur boxer. He had joined the SS in 1931 and the SD in 1934. Naujocks' first SD mission took place in early 1935, when he, together with a mistress and a colleague in a badly bungled operation, assassinated a German radio engineer and destroyed a radio transmitter broadcasting anti-Nazi propaganda from Czechoslovakia. His orders had been to kidnap the engineer and bring him to Berlin. Heydrich was initially furious, roaring about behavior suitable only for a gangster movie, but later put his trust in Naujocks again. Naujocks was therefore the man who directly started World War II by leading a fake Polish attack on the German radio station at Gleiwitz, near the Polish border, on 31 August 1939. On the air, Naujocks then personally read a brief, threatening statement in heavily accented Polish. This message was then broadcast to the German nation. Upon leaving, Naujocks and his men also left freshly killed concentration camp prisoners dressed in Polish military uniforms on the scene. This provocation provided Hitler with the excuse he needed to launch the war against Poland.

Naujocks' later career was varied. By the end of 1939, he was appointed head of Section VI F of the External SD (SD-Ausland), the foreign intelligence service (successor of the old Section III of the SD). There he was put in charge of Subsection VI W 1, which fabricated false passports, identity cards, and foreign bank notes for SD agents operating abroad. He also took part in other minor escapades, until Heydrich dismissed him from the SD in 1941 for disputing orders. Naujocks was accordingly transferred to the Waffen-SS and in 1943 sent to the eastern front.

There were also numerous amateurs among the less murderous SD intelligence agents and analysts. Foreign intelligence, for instance, was in the hands of the inexperienced SS-Oberführer (brigadier) Heinz Jost, director of Section III 2 (foreign intelligence services). The Jewish section of the SD ran its own intelligence organization in the Middle East, which was busy collecting intelligence on the conflict between Jews and Arabs in Palestine. This organization had close and friendly contacts with the Zionist organization Haganah.

Eventually, the External SD, as the foreign intelligence section later became known, controlled several thousands of agents abroad, for instance the famous Elyesa Bazna, code-named "Cicero," who was controlled by SS-Sturmbannführer (Major) Moyzisch at Ankara, Turkey. Bazna was a valet of the British ambassador in Ankara and from autumn 1943 to spring 1944 provided copies of secret documents from the safe in the ambassador's bedroom. These documents proved the stability of the alliance against Germany. Many foreign SD agents were unknowing agents, who did not know the true destination of their reports.

All these foreign activities caused the same kind of problems that the SD had with the Gestapo inside Germany, but this time with the Abwehr of Admiral Wilhelm Canaris outside Germany. Again, agents interfered with each other's operations, only this time abroad.

On 21 December 1936 a compromise of sorts was reached. Admiral Canaris and the judicial expert of the Gestapo, Dr. Best, signed an agreement appropriately code-named "The Ten Commandments." The military intelligence service was to deal with foreign intelligence and the exposure of foreign spies in Germany. The Gestapo, however, retained the duty to investigate any kind of treason, as well as the right to continue any other investigation that had originated in a treason investigation. The Gestapo also retained all executive powers, something that the military intelligence service lacked in times of peace. Only in wartime did the military set up a military police. Only the Gestapo was allowed to make an arrest, even if the military intelligence service had located and exposed a foreign spy.

The SD, however, was not technically part of this agreement. Indeed, on 11 November 1938, a decree proclaimed the SD the intelligence service of both the party and the state. Its principal duty was to aid the SIPO (that is, the Gestapo and the Kripo). The decree mentioned no limits on activities abroad. Meanwhile, the amateurish agents of the SD continued to expand their activities. The military increasingly resented the activities of the SD, and even the top leaders of the military agreed about one thing: the SD—being the police force of the SS—should not have any authority over the military.

THE RSHA

Neither Reichsführer-SS Himmler nor Heydrich wanted to admit defeat by the military. During the summer of 1938 Himmler initiated a grand project. He wanted to fuse the SS and the police into one unified organization, with the responsibility to protect the entire Nazi Reich.

Heydrich therefore began preparations for joining the SD and the Gestapo into one Reich security service, which caused problems. In the middle of 1935, when Heydrich split the role of the SD in two, party organ and intelligence organization, only 244 out of 607 Gestapo officials belonged to the SS. Even in 1939 only 3,000 of the 20,000 officials of the Gestapo were members of the SS. The fact remained that the SD had remained a party organization, while the Gestapo had remained a state organ. The executive organ—the Gestapo—had continued to be run as a state police organization, while the intelligence service—the SD—had continued to be run by party intellectuals with little, if any, administrative background. Although involved in many fields of inquiries, the SD did not have the wide powers of the Gestapo, and the organization never did have such a stable foundation within the formal structure of the Reich, as had the Gestapo.

The result of Himmler's grand plan was a not very efficient compromise. On 27 September 1939, the various organizations were fused into the new government authority known as the Reichssicherheitshauptamt (RSHA, or Reich Main Security Office), under the directorship of Reinhard Heydrich. The office of the new organ remained at 8 Prinz Albrechtstrasse, in the offices also occupied by Heydrich and the Gestapo. The new office was forbidden to ever appear in public with this name, and it was not ever to be used in a letterhead. To the public, there was only the "Director of the SIPO and the SD." Inside the organization, the state and party organs remained separate. To further confuse the situation, the members of the new organization, whether belonging to the Gestapo or the Kripo, used uniforms with the distinctive insignia of the SD. Every member was thereby—at least in theory—integrated into the SS.

The formation of the RSHA did not mean that the various components of the security apparatus—the Gestapo, the SD, the Kripo, and so on—were united into one organization. Each component remained separated from the others, and the rivalries of the various services remained. It was only at the top level that any unification was visible. Heydrich remained head of the RSHA, as he had been the real head of most of the services that went into the RSHA. (For a detailed description of the internal organization of the RSHA, see Appendix A following this historical introduction.)

The complex nature of the RSHA and its numerous departments and services made it necessary to familiarize all its agents and employees with the total structure. According to a circular dated 18 May 1940, written by Heydrich, every young agent who entered the RSHA had to serve at various stages in the different services. A young Nazi fresh out of the SS or a university at first spent four months at the Kripo, where he learned the fundamentals of police work and the basic notions of police science. Then he spent three months at the SD, followed by three months at the Gestapo. Only then, after he had acquired a familiarity with the needs and abilities of the various parts of the organization, was he permanently assigned to one of the services.

Within the Gestapo and the SD security was strict. In every office was posted the following admonition: "You must only know what is essential for your work; whatever you learn, you must keep it to yourself." More than one Gestapo official or collaborator was shot for breaking this rule. One official was even executed for merely telling a colleague from another section about the work he had been involved in.

Information was classified in one of four different security classes: (1) secret, (2) very secret, (3) secret information exclusively for the headquarters command, and (4) secret information of the Reich. The last category was only disseminated to the top leaders of the Third Reich.

Hitler had himself signed an order, the "Order No. 1 to all military and civilian authorities" of 23 May 1939. This order stipulated the following:

1. Nobody must know secret information that is not his proper concern.
2. Nobody must know more than what is strictly necessary for the execution of his duty.
3. Nobody must learn more about the obligations demanded by his duty than is strictly necessary.

4. Nobody must tell any subordinate more than he needs to know to accomplish an ordered task, or tell him before this is strictly necessary.

The strict security, "the patriotic duty of silence" as it was sometimes called, could not help but reduce the effectiveness of the security and intelligence organs of the RSHA. Subordinates were not provided with sufficient background information to act speedily and independently. Excessive compartmentalization was another, closely related problem. Nobody really knew what any other section was working on. The results were lost opportunities and interference in each other's operations. Furthermore, information had to pass from the lowest level in one RSHA hierarchy, the information-collecting one, to its top; from there it was transferred to another hierarchy, the executive one, and then being passed down again to the end user in this hierarchy. As reports were not only passed on but also rewritten at several levels, the final product often bore little resemblance to the original report. The RSHA was an organization not only of all-powerful secret police officers but—to an even higher extent—of petty-minded bureaucrats, each jealously guarding his personal domain and privileges.

During the war, a new security measure was introduced: anyone who was privy to state secrets was not, under any pretext, to participate in operations that risked his falling into the hands of the enemy. According to this, regular members of the Gestapo and the SD were never sent to the front. This, on the other hand, prevented them from directly learning the truth about the enemy and wartime conditions. (An exception to this rule was, amazingly, the RSHA chief and skilled pilot Reinhard Heydrich, who spent the first six weeks of the campaign against the Soviet Union flying with the Luftwaffe, in a show of characteristic bravado.)

The various security measures hampered the work of the RSHA in many ways. By the end of the war, the situation in the intelligence services had become disastrous. Everybody in the military high command and the upper levels of the RSHA knew that the war was lost, but nobody was allowed to mention it in an intelligence report. Hitler, becoming increasingly mad, reminded his officers of their duty of silence. Defeat was an inadmissible topic of conversation. Hitler even decreed that "whoever disobeys this order will be shot, regardless of rank or position, and his family will be imprisoned." Nobody dared to speak out, and Germany continued its slide into total defeat.

Or almost nobody. The one voice that stubbornly continued to report the true situation of the Reich and the war belonged to SS-Brigadeführer (Major General) Otto Ohlendorf, the youthful leader of the Internal SD. One of the most controversial leaders of the SD, Ohlendorf remains a real enigma. The intellectual Ohlendorf was a true believer, who could not understand that his party and Führer did not function better than they did. He murdered 70,000 Jews and reasoned that the rule of law had as its only purpose to secure the state and the nation, at the expense of the individual. He also argued, for instance, that the police was duty bound to "correct" any too-mild sentence or acquittal in court. Despite this, he also argued, in vain, against the primitive brutality of the Nazi regime. Himmler called Ohlendorf an "intelligent monstrosity."

As chief of the Internal SD, the efficient and conscientious Ohlendorf established a regular opinion research procedure in which the daily attitudes of the German people were revealed. These findings he published every two or three weeks in a special newsletter, *Meldungen aus dem Reich* (*Reports from the Reich*), distributed to the top Nazi leaders.

Ohlendorf's staff did a good job. It soon noted the megalomania of most top party officials, the waste of resources in the local party administrations, and the valueless and implausibility of the party propaganda. The German people, although obeying orders and staying enchanted by the Führer, did no longer trust the Nazi party. The idealistic Ohlendorf was shocked. He, who truly believed in the Nazi ideology, saw here with his own eyes, in the reports from his own trusted analysts, that something was wrong. His response was as split as his personality. At the same time that he advised high SD leaders that the situation could not be as bad as his own newsletter suggested, he stubbornly continued to publish his newsletter. As the war situation went from bad to worse, Ohlendorf got a reputation in the top Nazi leadership as a sick-minded pessimist. He especially clashed with Dr. Josef Goebbels, the minister for propaganda.

As Ohlendorf's newsletter was distributed more widely among party leaders, the criticism against him grew. As the reputation of Ohlendorf—and through him the entire SD—sank among loyal party members, the party initiated a fierce reaction against the SD. The party had long resented the way the SD supervised the party's internal affairs, and the increasingly negative reports from Ohlendorf proved a convenient excuse. During the autumn of 1942 and the spring of 1943, high party officials often demanded that the SD no longer be allowed to meddle in the internal affairs of the Nazi Party.

Reichsführer-SS Heinrich Himmler soon gave in to the demands. On 18 March 1943, he assured top party leaders that the SD had strict orders not to concern itself with the internal affairs of the party. Himmler's decision was undoubtedly eased by the fact that he resented the intellectual Ohlendorf, who had the unfortunate habit of lecturing the Reichsführer-SS on the ideals of the Nazi ideology.

Ohlendorf, however, stubbornly continued to publish *Meldungen aus dem Reich,* although Himmler often tore the newsletter apart and then returned the torn pages directly to Ohlendorf. Himmler also refused to pass a copy of the newsletter to the Führer. *Meldungen aus dem Reich* was simply too pessimistic to bother the Führer, who might be disturbed by its contents, Himmler reasoned. The truth was, as always, of less importance than the creative talent and dynamic will of the Führer.

Various other means to curb the readership and circulation of the newsletter, which by now had spread throughout the ministries and even into the state-owned business enterprises throughout Germany, were also taken. By the summer of 1943 the number of readers was drastically cut by decree. One year later the newsletter was completely banned. This did not prevent the rot that had already set in. The party had lost confidence in the SD. An increasing number of party organizations cut all connections with the SD and even forbade its members and employees from having any contacts whatsoever with the SD.

The SD had gradually lost its power and significance. As a party organization, funding became ever more difficult to arrange, and no national funding was awarded, except to certain individual sections. By 1944, when Ohlendorf's newsletter was banned, Himmler ordered the SD to surrender most of its functions. Only foreign intelligence and its role as a home for numerous special forces commands and the special commands responsible for political terror and genocide remained to the remnants of the SD.

Otto Ohlendorf had lost the struggle. Not only was he forbidden to continue reporting the truth, but his actions had destroyed the entire credibility of the intelligence service. His only consolation was a promotion to SS-Gruppenführer (lieutenant general) in November 1944.

The only one who gained anything from Ohlendorf's unlikely struggle between truth and propaganda, between Nazi ideology and service to the state, was Heinrich Müller of the Gestapo. For the Gestapo, human beings were merely a resource to be put to proper use by the omnipotent state. As the property of the state, the citizens existed only to be exploited as workers, taxpayers, and soldiers. Truth did not matter, only obedience to the state. And this was what the Gestapo excelled in.

THE GESTAPO GOES ABROAD

The war naturally brought major changes to the organization and activities of the RSHA. After the initial, speedy successes of the war, the RSHA found itself in the position to set up its services on site in several occupied territories. These services replicated the functions of the central RSHA. In this way the name and terror of the Gestapo spread throughout Europe. Strictly speaking, however, these local services were not part of the RSHA, even though they reported to the central RSHA headquarters in Berlin.

The Gestapo was indeed the service that profited most from the war. By the end of 1941 and the beginning of 1942, Heinrich Müller expanded the activities and resources of the Gestapo by also beginning operations in foreign countries that were not occupied by German forces. These countries had formerly been the preserve of the External SD and the military intelligence service. From then on, the Gestapo began operations abroad with the excuse of facilitating its work in counterintelligence. Müller thus obtained the right to correspond directly with the German police attachés who were officially or clandestinely installed abroad, and to demand information from and issue orders to them without first passing through the External SD.

At the time of the most intense activities of the local services of the RSHA, in spring of 1944, its external services and branches comprised 25 principal posts, 65 posts, and monitors at the 300 principal posts and 850 posts of the border police, which was part of the Gestapo. By the end of 1944, the total number of

Gestapo personnel—inside and outside the Reich—was from 35,000 to 40,000 permanent members (according to Ernst Kaltenbrunner, the last director of the RSHA, at the Nuremberg war crimes tribunal) to possibly 45,000 to 50,000 (figures advanced by the prosecution at the same time). These numbers, however, do not include the vast numbers of informers nor the auxiliaries recruited in the occupied territories. By the second half of 1944, the Gestapo absorbed a number of other security services, which increased its manpower further.

Himmler, in an order dated 1 October 1944, even passed under the control of the Gestapo the 54,000 agents of the Border Customs Police (Zollgrenzschutz), which until then had been controlled the Ministry of Finance. (Most of them continued to be paid by the Ministry of Finance, and shortly before the end of the war they were all reintegrated under its control. They played no significant role in the Gestapo.)

Minor changes also took place. By the end of 1941, a special section for prisoners of war was established within the Gestapo, led by SS-Hauptsturmführer (Captain) Franz Königshaus of Section IV A. At the beginning of 1943, this section was attached to the Subsection IV B 2 a, under the command of SS-Sturmbannführer (Major) Hans-Helmuth Wolf. The objectives were, among other things, to locate those prisoners who might be used in the occupied territories and to eliminate those who might pose a threat to the Nazi regime.

The outbreak of war also reinforced the power of the SS. At the beginning of 1940, Heinrich Himmler decreed that for the duration of the war, all German police forces were placed under the jurisdiction of the SS instead of the judicial tribunals. This removed the police from any kind of state control, and Himmler, in his capacity as Reichsführer-SS, could rule arbitrarily whenever he so desired. Eventually, in August 1943, Himmler also became minister of the interior of the Reich, gaining jurisdiction over the courts and the civil service.

By then, Dr. Werner Best realized that there was no longer any place for lawyers in the organization, which he had so eagerly helped to build. He resigned in May 1940, spending the rest of the war with administrative duties, first in occupied France and then Denmark.

Meanwhile, the professional police officers of the Gestapo worked on, but as Dr. Best had realized, nobody was any longer interested in the legal basis of operations.

In the occupied countries, too, the local branches of, especially, the Gestapo grew fast in power and importance. Because the peacetime military did not have any police force, the military command at the outbreak of war had been forced to turn to the Gestapo. The military leaders needed assistance in forming a military police, and despite the rivalry between the military and the security organs, whom else could they ask? The Gestapo happily agreed to supply the necessary cadres for the new Geheime Feldpolizei (GFP), or Secret Field Police, which was now formed and placed under the High Command of the Military Forces (Oberkommando der Wehrmacht, OKW). Technically, the GFP was independent of the RSHA.

In France and Belgium, the GFP originally counted 2,500 men. There was, however, also a small team from the SIPO (the Gestapo and the Kripo) led by 30-year-old SS-Obersturmbannführer (Lieutenant Colonel) Dr. Helmuth Knochen. Originally only a *Sonderkommando* (autonomous command with a special mission) 10 to 20 men strong, the small SIPO force had to serve in army GFP uniforms and travel in army vehicles. Their first job was to enter the police headquarters in Paris and retrieve all files on Germans living in France, Jews, and a certain number of politicians and others hostile to the Nazi regime.

Knochen's small force established a fruitful cooperation with the French police. A number of auxiliary French organizations were soon formed, to help the SIPO in its work. The military GFP was also brought under the control of the SIPO force, when Reichsführer-SS Heinrich Himmler on 7 May 1942 sent his personal representative, then SS-Brigadeführer (Major General) but soon Gruppenführer and Lieutenant General of Police Karl Oberg to take command of all SS and police forces in occupied France. The RSHA of Reinhard Heydrich soon absorbed the GFP in the occupied countries. Presumably this was his plan already when he sent the small SIPO force to France. No fewer than 5,000 members of the GFP became Gestapo informers.

Karl Heinz Hoffmann, the ruthless old political advisor of the Gestapo in Düsseldorf and the former head of the Subsection IV D 4 (the occupied Western territories: the Netherlands, Belgium, and France) of the

Gestapo, was appointed director of the SIPO in the Netherlands. Finally, he became joint director of the Gestapo in Denmark. Hoffmann immediately devoted himself to the struggle against the Resistance. Hoffmann was also one of the men responsible for the wording of *Nacht und Nebel* (*Night and Fog*), the order to execute hostages for every German soldier killed by the Resistance. This order caused the deaths of thousands of innocent civilians, most of them picked at random and executed to scare the surviving population away from the Resistance. The ones who were not executed at once were secretly transported to concentration camps in Germany, and no information as to their whereabouts or fate was ever given to their relatives.

East of Germany, the situation was different. There the military command had already from the beginning relinquished its right to maintain an independent military police. All police forces were controlled by the local branches of the Gestapo and the SS.

In Poland, Reinhard Heydrich had from the outbreak of war relied on five Sonderkommandos. These consisted of approximately 2,500 men, each wearing the SS field uniform but with the SD insignia on the left arm. Their mission was to destroy the Polish leadership: nobles, priests, teachers, officials, and intellectuals. Tens of thousands of the latter were killed during the war and immediately after the occupation of the country. As soon as the country was occupied, the five Sonderkommandos were reorganized as permanent Gestapo branches, one in each of the five districts, led by commanders from the SIPO and the SD. The branch commanders were subordinated to the director of the SIPO and the SD in Poland, who in turn answered to the RSHA.

In addition, there was also the ORPO, which had been mobilized at the outbreak of war, and numerous police battalions had taken part in the war against Poland and as members of the Sonderkommandos. Each Polish district also had a chief of ORPO, answerable to the chief of ORPO in Poland, who in turn was controlled by the central ORPO command in Germany.

By late 1941, Reinhard Heydrich considered the war a success for his security forces, both in the Reich and in the occupied countries. He prepared to reap the fruits of his hard work. Heydrich left his headquarters at Prinz Albrechtstrasse and moved to Prague, where he had been appointed deputy Reich protector of Bohemia and Moravia.

Heydrich had a plan to pacify his new domain. First he adopted a temporary policy of sudden repression and extermination of dissidents. Then he went on to win over the Czech workers and peasants by improving social conditions. Heydrich was successful, too successful, reasoned the Czech exile government in London and its British protectors. If Heydrich continued to bring improved condition to the Czech people, then there would no longer any need for an exile government, a horrifying thought. And the British much preferred a Czechoslovakia hostile to the Germans, and a situation that demanded troops and other resources that Germany sorely needed elsewhere.

The solution London chose was simple. Heydrich had to die.

The assassination of Heydrich appeared to be a simple mission. Confident in his success among the Czechs, Heydrich, in yet another display of bravado, had abandoned normal security procedures and driven about in an easily recognized open car, without armed escort. London prepared two former Czechoslovak soldiers, Jan Kubis and Josef Gabcik, for the mission. After completing their training in England, the two were parachuted into Czechoslovakia. Two other Czech resistance men joined them, and the small team prepared to ambush Heydrich on his daily drive from his residence to his Prague office.

On the chosen morning, 27 May 1942, the four men, armed with submachine guns and hand grenades, awaited the arrival of Heydrich. At first everything appeared to go wrong. Heydrich was late, and the four men grew increasingly nervous. Then, finally, they saw Heydrich's green Mercedes approaching. Gabcik jumped out in front of Heydrich's car and opened fire with his submachine gun, which promptly jammed. Kubis then threw a hand grenade at the green Mercedes, which exploded, damaging the front of the car. The car, sliding and braking, soon stopped. The two Czechs closed in for the final kill.

Heydrich, however, now rushed out of the car, shouting and firing his pistol at the two Czechs. Amazed, the two Czechs fled, pursued on the street by the enraged Heydrich. Both made their escape, however, and it was soon revealed that Heydrich, after all, was gravely wounded from the explosion. The doctors could do nothing, and Reinhard Heydrich died of his wounds on 4 June 1942.

The enraged German security forces launched a campaign of savage reprisals throughout the country. More than a thousand Czechs were executed, including all male inhabitants of the village Lidice, and 10,000 were arrested. Both Kubis and Gabcik died during the reprisals. The killings turned the national spirit against the Germans, and the dormant Czech resistance organization soon overflowed with recruits, eager to fight the Germans. London had succeeded in its mission.

After the assassination of Heydrich, Himmler began a search for a new chief of the RSHA. On 30 January 1943, he appointed the SS-Gruppenführer and Lieutenant General of Police Dr. Ernst Kaltenbrunner (1903–1946), a nearly seven-foot-tall Austrian veteran Nazi from the same region as Hitler. Whereas Heydrich had been ruthless and cold, but calculating and efficient, Kaltenbrunner, an excitable alcoholic and chain-smoker, was merely brutal. He was not the man to create an intelligence and security service, but he was quite capable of running it and providing victims for the gas chambers and firing squads. Kaltenbrunner's speciality was destruction and brutal repression.

From August 1943, the last restraining voices within the German police were silenced. When Heinrich Himmler also was appointed Reich minister of the interior, he transferred the few remaining control possibilities of the police from this ministry to the RSHA. This especially affected the ORPO, whose headquarters no longer had any authority over its own forces. Himmler also dissolved the ORPO section for administration and judicial affairs and removed its troublesome and legalistic head, Werner Bracht, from his post. This section was instead reorganized as an office for financial administration, headed by veteran Nazi SS-Gruppenführer (Lieutenant General) August Frank.

At about this time, Himmler's rival Kurt Daluege, chief of the ORPO, fell ill. In the summer of 1944, most remaining responsibilities of the ORPO were also transferred to the RSHA.

By the end of 1944, the Gestapo also absorbed a part, the Abwehr. The military counterintelligence service of Admiral Wilhelm Canaris had for a long time been a rival to the RSHA, but the prestige of Canaris had prevented the Gestapo from directly controlling this military organization. Not only did the Gestapo resent this lack of control; there was also a hard core of anti-Nazi officials in Abwehr, some of whom plotted actively against Hitler and Nazi rule.

The military intelligence service, the Amt Ausland Nachrichten und Abwehr (Foreign Intelligence and Defense Office), formed one of the five directorates of the OKW. The military intelligence service was divided into two main offices, the Amtsgruppe Ausland, which dealt with general information, important but not secret, and cooperated with the Ministry of Foreign Affairs, and the Abwehr Amt, which essentially was the secret military intelligence service. The Abwehr Amt was divided into four *Abteilungen* (departments), of which Abteilung I (secret reports), Abteilung II (sabotage), and Abteilung III (foreign agents) were the most important.

In spring 1943, the Gestapo had enough evidence of anti-Nazi activities to make a first strike against the military counterintelligence service. Several leading officials were arrested. Yet more arrests followed in January 1944. Other members of the military counterintelligence, serving abroad, defected to the Allies. This sealed the fate of the service as an independent organization.

By February, Hitler grew tired of the military counterintelligence and ordered all remaining parts of it to be subordinated to and assimilated with the SD. The Abwehr was formally dissolved on 14 February 1944. Admiral Canaris was replaced by SS-Brigadeführer (Major General) Walter Schellenberg, chief of Amt VI (External SD) of the RSHA (by the end of 1939, Schellenberg had been jumped from the rank of Sturmbannführer, or major, to Brigadeführer). Unlike Müller of the Gestapo, Schellenberg was a friend of Canaris. Although Schellenberg could not avoid arresting Canaris, he dismantled the military counterintelligence service with care. When possible, he allowed its personnel to remain in their old posts.

Canaris, too, was eventually allowed to return to Berlin to take up new duties as head of the military special staff for trade war and economic warfare. Schellenberg, it appears, secretly sympathized with the German resistance groups that wanted to replace Hitler and arrange a separate peace with the Western Allies. At the end of the war, however, in April 1945, Canaris was executed.

The central administration of the Amt Ausland Nachrichten und Abwehr was dissolved. Amtsgruppe Ausland became attached to the Wehrmachtsführungstab (staff for the leading of operations) of the OKW.

The four foreign sections of Abwehr Amt were absorbed by the RSHA as a supplementary Amt, called the Militärisches Amt (Military Office), or simply Mil Amt. Its new head became Colonel Georg Hansen, the former chief of Abteilung I of the Abwehr, who now replaced Canaris' friend Pieckenbrock in this position. Abteilungen I and II (sabotage) were merged as a new military counterintelligence service. Abteilung III was shared between Mil Amt and the Gestapo.

Colonel Hansen, however, remained part of the resistance against Hitler. After the July 1944 attempt on Hitler's life Hansen was arrested and hanged.

Hansen's arrest and execution was the end of the Mil Amt. The service was split between the Gestapo (Amt IV) and the External SD (Amt VI) of the RSHA. The Gestapo assumed control of the sections dealing with foreign intelligence, counterintelligence, parachutists, and sabotage. The External SD received the military intelligence section. Each subsection of the old Mil Amt was assigned to a subsection within the Gestapo or SD, where it received the same administrative assignation as its parent subsection, but followed by the suffix *Mil.*

The new ex-military subsections of the SD were eventually (on 15 January 1945) put under the control of Helmuth Knochen, who by then had been recalled from France to be demoted to a simple Waffen-SS private and then been happily reinstated with some of his old standing.

The RSHA leaders differed in their opinions on one important question. If the war were to end in defeat, as seemed more and more likely, how should the RSHA and the party react? For Ernst Kaltenbrunner and Heinrich Müller, the answer was easy: the war could not be lost, and any opposition to the regime must be ruthlessly suppressed. Other RSHA leaders could see more clearly. The moral dilemma of Otto Ohlendorf regarding his newsletter has already been discussed. Walter Schellenberg, since June 1941 deputy chief and from 1942 the formal head of the External SD, was also able to see that the war was not going well. From 1944, when he was appointed head of the united SD and the remnants of the military intelligence service, he went into secret negotiations with the Allies about surrendering the German armies in the west.

Another man was also, it seems, involved in the resistance against Hitler. The ambitious SS-Obergruppenführer (General) Arthur Nebe, head of the Kripo, responsible for the deaths of at least 46,000 Jews, for obscure reasons involved himself in the July 1944 plot against Hitler. After the failed assassination of Hitler, Nebe—although not under suspicion—went into hiding. He was eventually betrayed by a rejected mistress and apparently hanged in Berlin in March 1945.

Even Heinrich Himmler, the Reichsführer-SS, by the end of the war dreamed of creating a separate peace with the Allies in the West. He ordered the mass extermination of Jews to be stopped and even planned to set up his own government.

Among the security officers who plotted against Hitler, one man was missing. The one man who had both the power, the ability, and—according to testimonies from his close associates after the war—the outspoken conviction to successfully eliminate Adolf Hitler should the Führer become a liability to the Reich did not take part in the July 1944 plot. He was already dead. Reinhard Heydrich, assassinated upon orders from London, may paradoxically have been the only one who could have ended the war in 1944.

THE GESTAPO AND THE JEWS

The man in charge of the Jewish section of the Gestapo was Adolf Eichmann (1906–1962). A boyhood friend of Ernst Kaltenbrunner, the young Eichmann drifted through many jobs before he finally found a home in the Nazi Party. In September 1934, he found an open position in the SD. There Eichmann's bureaucratic talents came into full flower. By the beginning of 1935, he was in charge of Section II 112, responsible for "Jewish Questions" at the SD headquarters in Berlin. Eichmann displayed a remarkable talent for this work, and he was soon regarded as an authority on the problem. He especially concerned himself with the Zionist movement, learned some Hebrew and Yiddish, and even briefly visited Palestine in 1937.

Eichmann was the perfect bureaucrat. He obeyed any order and had few, if any, personal opinions. He was, though, concerned about the development of a future Jewish state. Eichmann saw a danger in such a state, as it could turn into a potential enemy of Germany.

The Nazi Party was at first divided in its opinion on what to do with the Jews. No Nazi liked the Jews—and it must be remembered that this feeling was widespread not only in Germany, but also in most parts of Central and Eastern Europe—but few, if any, were yet so radical as to wish for their total destruction. Heinrich Himmler, the Reichsführer-SS, thought that the Jews should be persuaded to emigrate. For this purpose, he supported the Zionist movement in its wish to set up a Jewish state in Palestine.

The Nazi Party, therefore, set up an Office for Jewish Emigration under the control of Heinrich Müller of the Gestapo. From August 1938, Eichmann was connected to this office, and in October 1939 he replaced Heinrich Müller as its head.

In December 1939, Eichmann was transferred to Amt IV (Gestapo) of the RSHA and put in charge of the Subsection for Jewish Affairs and Evacuation (Referat IV B 4). A new office was established at 116 Kurfürstenstrasse in Berlin. This was to be the office in charge of the coming destruction of the Jews. (Eventually, Eichmann was made the head of the entire Section IV B.)

The SS and RSHA leaders were still not unanimous in a wish to destroy the Jews. Even Eichmann himself still promoted emigration as the way to remove all Jews from the Reich. Himmler, too, who was disgusted by cruelty to animals and, unlike most Nazi leaders, never abused his powers for personal profit, remained in favor of emigration. A shift in emphasis had, however, taken place. What at first had been a more or less voluntary emigration was turned into a forced exodus and, finally, a deportation of Jews to special areas in Poland. It was still generally expected that these areas were mere staging areas for emigration. However, the fact that these areas were turned into what may be called open concentration camps naturally facilitated any kind of later decision.

In early March 1941, Hitler decided that in the planned war against the Soviet Union, the Germans were to exterminate all Jews and Communist leaders on Soviet territory. Soon this was expanded to include all Jews and all Soviet Communist Party and state officials. The task was to be executed by the GFP and special units of the SS and the police. In April 1941, Reinhard Heydrich called the high officers of the RSHA together and asked for volunteers; it was to be a "tough mission" and he needed "real men" who could follow orders.

It is unclear whether many of the RSHA officers knew what Hitler and his loyal subordinates Himmler and Heydrich really had in mind. Maybe they did, maybe not. In any case, only one volunteered: Arthur Nebe, head of the Kripo. Others had to be ordered to go. Nobody refused; this would have been a certain fall from favor. Besides, nobody may have known exactly what was going to happen. Although such a favorable view may seem unlikely, it remains possible when one considers the excessive secrecy within the RSHA.

In the end, several RSHA officers were chosen and sent to lead a number of special action groups (Einsatzgruppen) in the east. Of them, only the ambitious Nebe—promoted for his loyalty to SS-Gruppenführer (lieutenant general)—and the loyal Nazi Otto Ohlendorf followed the orders with apparent enthusiasm. Jews and Communists were systematically murdered throughout the occupied territories, often immediately following the army's occupation of a town.

The lower-ranking officers of these teams were hardly more enthusiastic than the majority of the top leaders. Most of them were intellectuals, who did not relish their new job. Among these were academics with double doctorates, lawyers, and even a priest and an opera singer. To get enough recruits, Heydrich had to borrow personnel from the ORPO and the Waffen-SS. An entire police battalion in Berlin was disbanded so that its personnel could be ordered to join.

By the end of May 1941, Heydrich had finally managed to appoint about 3,000 men to the special groups, divided into 120 sections. On 22 June 1941, the day after Hitler had begun the war against the Soviet Union, their task began.

The extermination groups murdered hundreds of thousands of Jews and real or imagined Communists, as well as numerous other civilians who happened to be in the wrong place at the wrong time. There is no need to go into individual details here, as the full scope and terror of the action have been amply detailed elsewhere. Of the Germans who were commanded to serve in these units, few enjoyed the work, and many contracted severe psychological problems. Yet nobody protested the order, unless one counts the passive protests of the many who applied to be transferred from their action group.

From this initial extermination of real and imagined enemies in the war, it was but a short step to the "final solution" of the Jewish question. In 1941, the loyal bureaucrat Eichmann was promoted to SS-Obersturmbannführer (lieutenant colonel). On 31 July 1941, Hermann Goering ordered Reinhard Heydrich to prepare for a total and final solution of the Jewish question in all German territories. On 20 January 1942, Heydrich convened the Wannsee Conference to discuss the way and means of implementing the "Final Solution of the European Jewish Question." In circumlocutory language, the participants of the conference finally decided and planned the destruction of the Jews. Eichmann's position as the authority on Jews was consolidated, and Reinhard Heydrich formally entrusted him with the implementation of the plan.

Eichmann, the loyal bureaucrat who "personally" had no ill will against the Jews, got to work with his customary zeal and industriousness. Supported by his immediate superior, Heinrich Müller, and, after the death of Heydrich, the RSHA chief, Ernst Kaltenbrunner, Eichmann devoted the rest of his career to the extermination of Jews. Soon the collection areas for Jewish emigrants in Poland turned into collection areas for the death camps. The full-scale extermination of Jews, Gypsies, and other minorities had begun.

THE END OF THE GESTAPO

The German surrender meant the end of the RSHA and the Gestapo. It also meant the end for the tens of thousands of officials who worked in these organizations.

The majority tried to escape. Wearing civilian clothes and often carrying forged identity papers, they spread throughout Europe, pretending never to have had anything to do with the security organs. Some succeeded, but most failed. Large numbers were co-opted by the Soviet intelligence service especially, but also by the secret services of the other Allied powers. Some were imprisoned, and upon release, especially in East Germany, often joined the Communist Party to bring their vast experience to government service.

Heinrich Müller, the chief of the Gestapo, simply disappeared at the beginning of May 1945. Many German officers who were prisoners of war in the Soviet Union later testified that Müller was seen in Moscow. According to Walter Schellenberg, Müller had profited from the affair of the Rote Kapelle (a Soviet spy ring the Gestapo had exposed) to establish contacts with Soviet agents. Müller presumably put himself at their service at the end of the war. According to these sources, Müller died in Moscow in 1948, at the relatively young age of 47. However, later sightings of him have been reported from Chile, where he was supposed to be staying with Nazi leader Martin Bormann.

Ernst Kaltenbrunner, head of the RSHA, was arrested together with Hermann Goering. Both were condemned to death in the Nuremberg war tribunal, and Kaltenbrunner was hanged at the end of 1946. Goering had taken poison a few hours earlier.

Heinrich Himmler, the Reichsführer-SS, committed suicide by swallowing a poison capsule hidden in his mouth after being arrested by British troops at the end of the war.

The attempted peace negotiations organized by **Walter Schellenberg**, head of the External SD, had led to nothing, and Schellenberg was condemned to prison after the war. Released by the end of 1950, he died in Italy less than two years later.

The loyal Nazi **Otto Ohlendorf**, because of his activities as leader of an Einsatzgruppe, was sentenced to death by the Nuremberg war crimes tribunal. He was hanged in 1951.

Dr. Werner Best, who had resigned from the RSHA in May 1940, spent the rest of the war involved in administrative duties, first in occupied France and then Denmark. In France, he was actively involved in the deportation of Jews; while in Denmark, he instead sabotaged Himmler's orders concerning the elimination of Jews. For these conflicting reasons, Dr. Best spent several years after the war in prison, alternating with years working as a lawyer. He was eventually freed from prison, if not from accusations, on medical grounds in 1972.

The ambitious **Arthur Nebe**, head of the Kripo, was apparently hanged already before the war ended. However, during the years 1956–60, he was sighted in Turin, Italy, and again in Ireland, where he was reported to be with the SS special forces leader and chief of section S of the External SD, Otto Skorzeny, the man who after the war founded the secret organization to help former SS members escape to South America. Israel sent a vengeance team after Nebe in Ireland, but by then he was gone, if he ever had been there.

Helmuth Knochen went into hiding after the war and was only arrested in early 1946. Eventually, in 1954, he was condemned to death in France, the country in which he had spent most of the war. He was never executed, however, and was eventually released in 1962, to face further trials in Germany. His superior in France, Karl Oberg, arrested in 1945, was sentenced to death in the same trial, but his sentence was also commuted to imprisonment. In 1965, the French president pardoned him for repatriation to Germany, where he died the same year.

The brawler turned killer, **Alfred Naujocks**, who had been dismissed from the SD in 1941 for disputing one of Heydrich's orders and transferred to the Waffen-SS, was in 1943 sent to the eastern front. He soon found his way back to the west, however, and after a few other assignments and assassinations deserted to the Americans by the end of 1944. At the end of the war, Naujocks found himself in a prisoner-of-war camp. This was not to his liking, so he promptly escaped and disappeared. It is rumored that he from then on, together with Otto Skorzeny, assisted fugitive Nazis in escaping to South America, and finally settled down as a businessman in Hamburg, where he appears to have died in 1960.

Adolf Eichmann, the man behind the implementation of the "final solution" of the Jewish question, was arrested and briefly put in an American internment camp. He escaped in 1946, however, and went into hiding. Eichmann eventually moved to South America in 1952. After traveling through Argentina, Brazil, and Paraguay for three years, he finally settled down in Buenos Aires in 1955 under the name of Ricardo Klement. His wife and two daughters joined him there, and he secured employment in a car factory owned by Mercedes-Benz. However, in 1960 a group of Israeli agents kidnapped him and flew him to Israel. There Eichmann was condemned to death and hanged in 1962.

Kurt Daluege, the chief of the ORPO had succeeded Reinhard Heydrich as deputy Reich protector of Bohemia and Moravia after the latter's death in 1942. He was executed by the Czechs in 1946.

Rudolf Diels, the first head of the Gestapo, had after his final dismissal from the Gestapo in April 1934, retained the protection of his life-long friend Hermann Goering. In May of the same year, he ended up as a local government official in Cologne. Six years later he briefly held a higher post in Hanover, but was dismissed after refusing to carry out an order to arrest Jews in the city. After the July 1944 plot against Hitler, Diels even ended up in a Gestapo prison. Goering, whose cousin Diels by then had married, always saved him from arrest or worse and rescued him from the prison. Diels survived the war and later worked in the West German Ministry of the Interior. He died in 1957 in a hunting accident.

All these deaths, however, were only the deaths of the leaders of the Gestapo and RSHA. The Gestapo itself died—piece by piece—in the final months of the war. The Gestapo was not so much the flesh-and-blood leaders, but the technical inventions and instruction manuals, the many card indexes, and, especially, the vast number of files that covered the most intimate secrets of millions of people, spread throughout centers and branch offices all over the Reich. Some of these perished in the flames of the incendiary bombs dropped on the German cities. Others were lost in burning cars while being transported from the invading Allied armies. A few survived and fell into the hands of the victors. Of these, some were used at the Nuremberg trials. Others simply vanished.

But did the Gestapo really die? The Gestapo—as an organization—was dead, but some of its vestiges were left in the hands of the victors, and the legacy of the Gestapo lived on. The files retrieved by the Soviet intelligence service were later, sometimes much later, used to blackmail prominent Germans into becoming Soviet spies. This much has been revealed after the collapse of the Soviet Union and the KGB. But what happened to the files that fell into the hands of the Allies in the West? It is well known that many Nazis were co-opted by the U.S. government. How about their technical knowledge of intelligence and security operations? How about any remaining files? How about their attitude to the proper utilization of a state's citizens? If there are those who know, they do not tell. The legacy of the Gestapo lives on, even now, and in places where one least expects to find it.

APPENDIX A

•

THE INTERNAL STRUCTURE OF THE NAZI SECURITY APPARATUS

Table 1

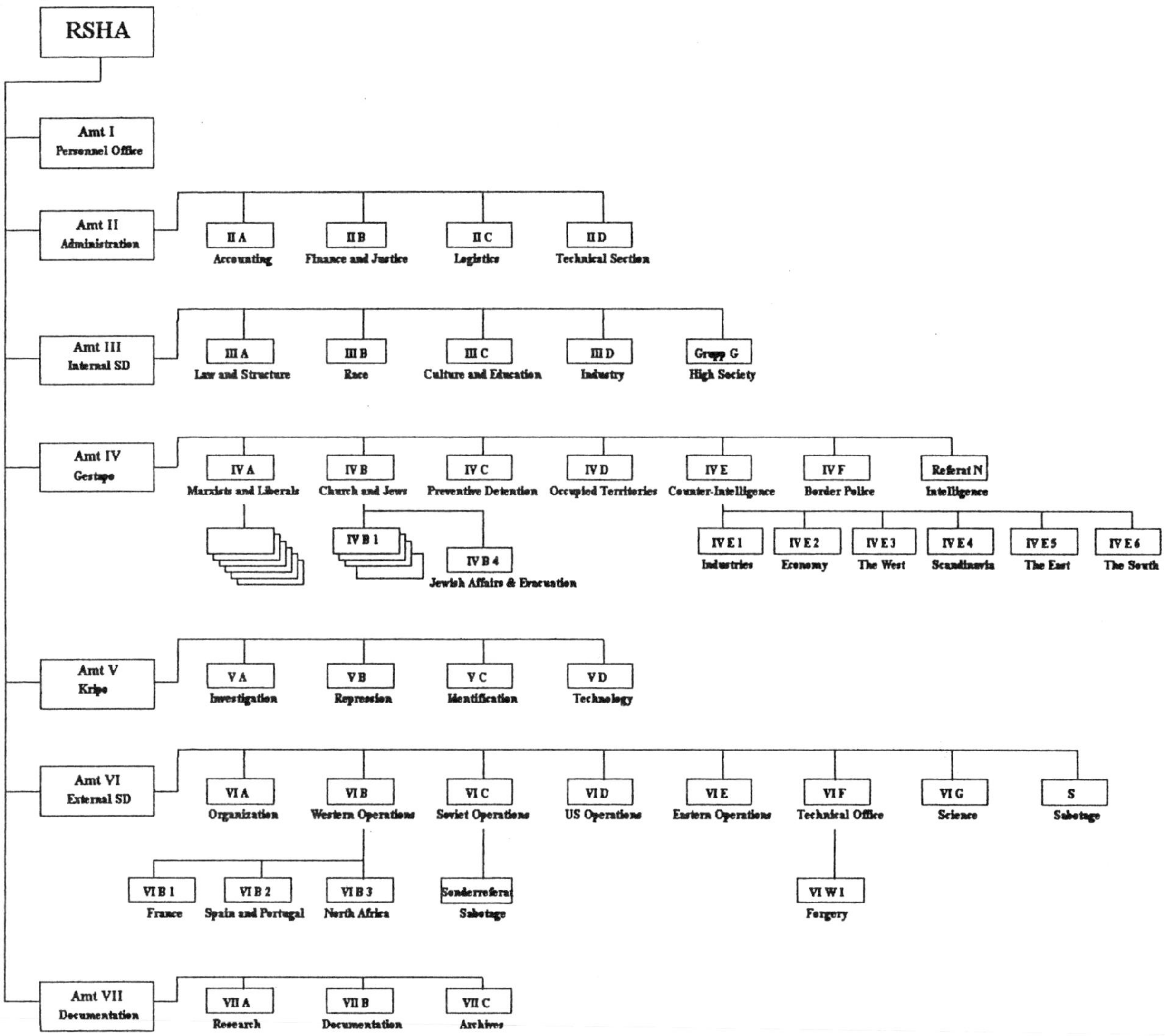

When fully developed, the Nazi security apparatus had grown into the Reichssicherheitshauptamt (RSHA), the Reich Main Security Office. The RSHA was divided into seven offices (plural *Ämter,* singular, *Amt*), the first two of which were purely administrative and had no local branches. The others were a curious mixture of state and party organs, of various origins in the German state or the Nazi Party. The last office (Amt VII) also had no local branches, even though it was not purely administrative.

Amt I: Personnel Office of the RSHA

Directors:

- Dr. Best, from the creation until 12 June 1940, when he was transferred to Paris
- Streckenbach, from July 1940 until the beginning of 1943
- Schultz, from the beginning of 1943 until November 1943
- Ehrlinger, from November 1943 until the surrender

Amt II: Administrative and Financial Matters

Divided into four sections:

II A: Offices, maintenance, salaries, and accounting

II B: Financial matters, liaison with the Ministry of Justice, responsibility for those in custody (with the exception of the ones in prisons and the camps), and transportation of same

II C: Logistics (for the active services of the SIPO and SD, see below)

II D: Technical section (especially for vehicles)

Directors:

- Dr. Best, from the creation until July 1940
- Nockemann
- Siegert
- Spacil

Amt III: Internal SD (Party Organ)

The internal Sicherheitsdienst (Secret Service), an active intelligence service, divided into five sections. The central office employed 300 to 400 agents. Amt III directed the vast network of informers operating in Germany.

III A: Matters concerning law and the structure of the Reich (Subsection III A 4, for instance, published *Meldungen aus dem Reich*, a newsletter on the general opinion and the attitudes of the population)

III B: Matters concerning the "ethnic community" of the Reich: ethnic minorities, race, public health

III C: Matters concerning culture, science, education, arts, and the press. Intelligence on religious affairs and places of worship (one section from this office was assigned to the Gestapo during the struggle against the churches on 12 May 1941)

III D: Matters concerning the economy, surveillance of the industries, food supply, commerce, etc.

Grupp (Group) G: Manipulation of "honorary agents" for intelligence collection in "high society"

Director:

- Otto Ohlendorf, from the creation to the end

Amt IV: Gestapo (State Organ)

The Geheime Staatspolizei (Secret State Police), an active service with executive powers (the right of arrest) in the political field, was divided into six sections. It collected intelligence on enemies of the Nazi regime and repression. The central office employed 1,500 agents.

IV A: Enemies of Nazi Germany: Marxists, Communists, reactionaries, and liberals; countersabotage and general security measures; Section IV A included six subsections

IV B: Political activities of the Catholic and Protestant churches, religious sects, Jews, and Freemasons; divided into five subsections, one of which (Subsection IV B 4, led by Adolf Eichmann) was in charge of the "final solution" of the Jewish question

IV C: Protective custody, preventive detentions; the press; party affairs; files

IV D: Occupied territories, foreign travelers in Germany; included Subsection IV D 4, in charge of the Western territories (the Netherlands, Belgium, and France) led by Karl Heinz Hoffmann

IV E: Counterintelligence; divided into six subsections:

—IV E 1: General matters of counterintelligence; counterintelligence in the factories of the Reich
—IV E 2: General matters of the economy
—IV E 3. Countries in the west
—IV E 4: The Scandinavian countries
—IV E 5: Countries in the east
—IV E 6: Countries in the south

IV F: Border Police (Grenzpolizei). Passports, identity cards, policing of foreigners; from 1941, there was also a supplementary independent section, Referat N, which supervised the centralization of intelligence.
The Gestapo went through several internal changes, but the organization and the duties remained the same until the end of the war.

Director:
• Heinrich Müller, from the beginning to the end

Amt V: Kripo (State Organ)
The Kriminalpolizei, an active service with executive power in the field of crime. Its central office employed 1,200 agents. Divided into four sections:
V A: Criminal Investigation Police and crime prevention measures
V B: Repression of crime and illegal activities
V C: Identification and research
V D: Institute of Criminal Technology of the Sicherheitspolizei (SIPO), or Security Police, the joint name of the Gestapo and the Kripo.

Directors:
• Arthur Nebe, until 20 July 1944 (he was later supposedly hanged because of his involvement in the July 1944 conspiracy against Hitler)
• Obergruppenführer (General) Panzinger, the former chief of Section IV A of the Gestapo, from this date to the end

Amt VI: External SD (Party Organ)
The SD-Ausland (External SD) was a foreign intelligence service. In this role, the duties of the organization overlapped those of the Abwehr, the military intelligence service. Its central office employed from 300 to 500 agents. Abroad, the office controlled several thousands of agents. At first divided into six sections, then into eight.
VI A: General organization of the intelligence service, and control of the work of the various sections within the SD (the latter objective only until 1941)
VI B: Intelligence operations in Western Europe. Three subsections:
—VI B 1: France
—VI B 2: Spain and Portugal
—VI B 3: North Africa

VI C: Intelligence operations in the area of influence of the Soviet Union. Included Subsection VI C 13 (the Arab section) and the Sonderreferat (Special Section), which was in charge of sabotage operations in the Soviet Union
VI D: Intelligence operations in the area of influence of the United States
VI E: Intelligence operations in Eastern Europe (except the Soviet-controlled countries)
VI F: Technical office for Amt VI (its director was the infamous killer Alfred Naujocks; included Subsection VI W 1, which fabricated forged foreign bank notes
VI G: Exploitation of scientific intelligence (created in 1942)
S: In charge of "material, moral, and political sabotage" and after April 1943 under the command of the famous special forces officer Otto Skorzeny (this section also created in 1942)

Directors:
• Heinz Jost, until the beginning of 1941 (when he was demoted in rank and sent to the eastern front as an ordinary soldier)
• Walter Schellenberg, (who was then only 32 years old and since 1939 had been director of Section VI A), from 1942 until the end

Amt VII: Documentation (Party Organ)
Ideological research among the enemies of the Nazi regime: Freemasonry; Judaism, the church, liberals, Marxists. Divided into three sections:
VII A: Central research and documentation
VII B: Exploitation of the documentation: syntheses of documentation, biographical notes, written commentaries
VII C: Central archives: classifying intelligence and information, card indexes; maintenance of the museum, library, and photographic archives

Appendix B

•

MAP OF THE GERMAN REICH

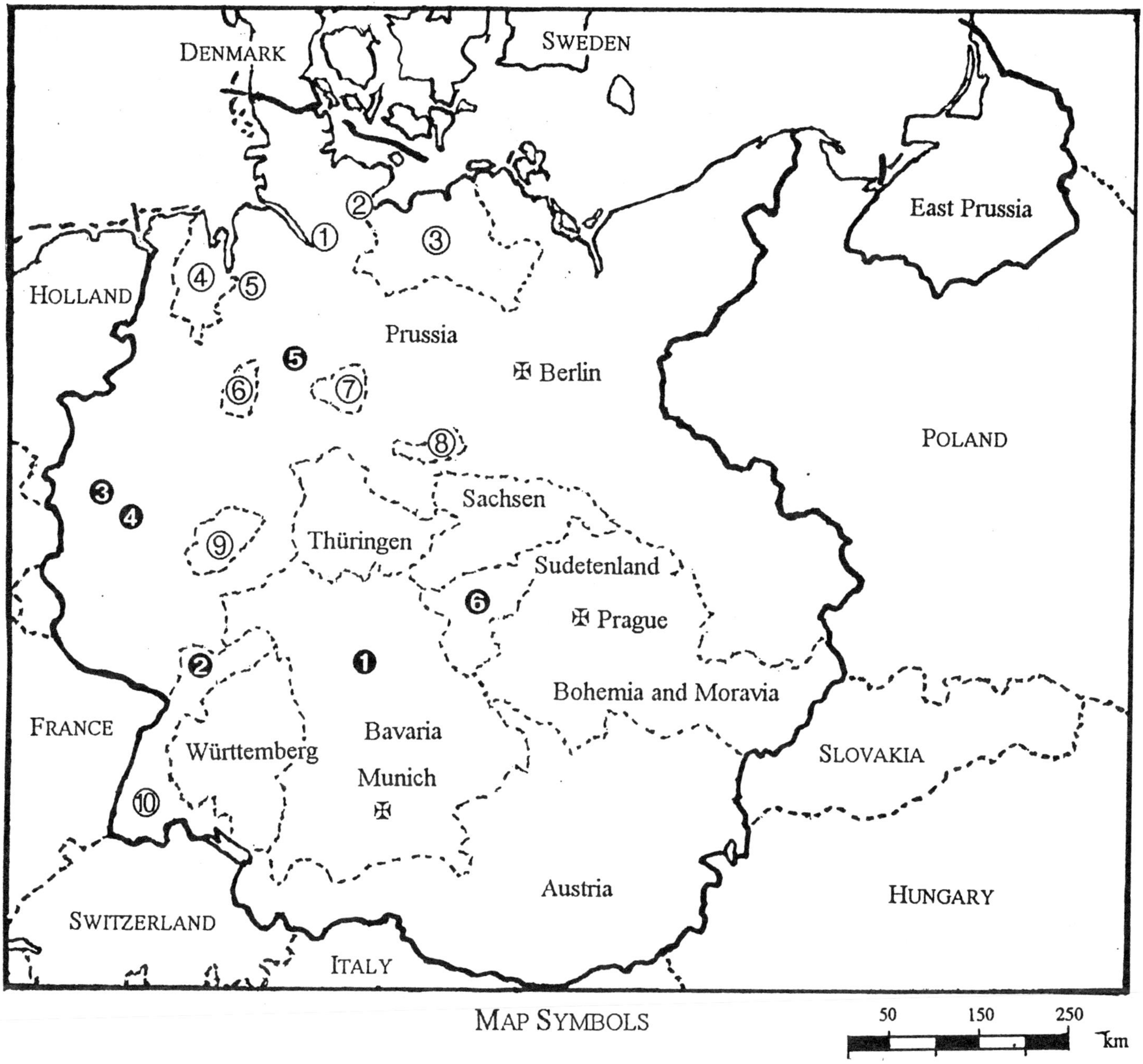

Map Symbols

Cities

❶	Nuremberg
❷	Heidelberg
❸	Cologne
❹	Bonn
❺	Hannover
❻	Karlsbad

States

①	Hamburg	⑥	Schaumburg-Lippe
②	Lübeck	⑦	Braunschweig
③	Mecklenburg-Schwerin	⑧	Anhalt
④	Oldenburg	⑨	Hessen
⑤	Bremen	⑩	Baden

Appendix C

FURTHER READING

Aronson, Shlomo. *The Beginnings of the Gestapo System.* New Brunswick: Rutgers University, 1969.

_________. *Reinhard Heydrich und die Frühgeschichte von Gestapo und SD.* Stuttgart, 1971.

Bazna, Elyesa (with Hans Nogly). *I Was Cicero.* New York, 1962.

Best, Werner. *Die deutsche Polizei.* Darmstadt, 1941.

Butler, Rupert. *An Illustrated History of the Gestapo.* Osceola, Wisconsin, 1993.

Crankshaw, Edward. *The Gestapo—Instrument of Tyranny.* London, 1956. Republished in New York, 1991.

Delarue, Jacques. *Histoire de la Gestapo.* Paris, 1962. Most recently published in English as *The Gestapo—A History of Horror.* New York, 1986.

Diels, Rudolf. *Lucifer ante portas: Es spricht der erste Chef der Gestapo.* Stuttgart, 1950.

Höhne, Heinz. *Der Orden unter dem Totenkopf—Die Geschichte der SS.* Hamburg, 1966. Most recently published in English as *The Order of the Death's Head—The Story of Hitler's SS.* New York, 1986.

Krausnick, Helmut and Hans-Heinrich Wilhelm. *Die Truppe des Weltanschauungskrieges—Die Einsatzgruppen der Sicherheitspolizei und des SD, 1938-1942.* Stuttgart, 1981.

Schellenberg, Walter. *The Schellenberg Memoirs.* London, 1957.

Wistrich, Robert. *Who's Who in Nazi Germany.* New York, 1984.

APPENDIX D

THE METRIC SYSTEM

As this handbook was written in Germany, all measurements follow the international metric system.

1 millimeter = 0.039 inch
1 centimeter = 0.39 inch
1 meter = 39.37 inches
10 meters = 32.81 feet

THE GESTAPO AND SS MANUAL

PART I

GENERAL INFORMATION

HOW TO READ MAPS

Different Types of Maps

The most important purpose of a map is to clearly show the significant features of the terrain, while neglecting, even ignoring, the unimportant ones. Railways, roads, and so on are prime examples of important features.

We need a map that always—wherever we are in the terrain—gives us the best possible advice. This wish is excellently fulfilled in the *Reichskarte*, the map of the German Reich, in the scale of 1:100,000. In this handbook, we will only use the Reichskarte, because:

(1) the Reichskarte is the most suitable map for our purpose, with map sheets available for all of Germany in the same scale,

(2) everybody who can understand the Reichskarte without problem can also read any other map, and

(3) there exists, except in greater scale, no other more detailed and better map.

The Scale

The map scale tells us how much a real section of the terrain has been decreased in size on the map. This fact enables us to locate this terrain section on the map.

In this way, for instance, the scale of 1:100,000 tells us that 1 centimeter on the map corresponds to 100,000 centimeters in the terrain. We all know that 100,000 centimeters equals 1,000 meters, or 1 kilometer. Therefore 1 centimeter on the map corresponds to 1 kilometer in the terrain. Any other scale is equally easy to understand. If a town map is made in a scale of 1:15,000, then 1 centimeter on the map equals 15,000 centimeters on the ground, which corresponds to 150 meters. Therefore 1 centimeter on the town map equals 150 meters on the ground.

The Map Symbols

Every sheet of the Reichskarte includes a symbol explanation. Study and learn these symbols conscientiously. The explanation of symbols is sufficient for understanding all symbols used on the map. Here we will merely say a few words about some of the symbols. If a railway or a road passes through a tunnel, its route in the tunnel is marked with lines (Fig. 1).

The embankments of roads and railways are indicated by short, fine lines at right angles to the road or railway. A ravine or cutting is indicated in a similar manner, but here the outer borders of the lines are indicated by an outline of the ravine or cutting.

When a road passes above a railway track, the symbol for the road continues across, while the symbol for the railway is interrupted. When a road passes under a railway, the opposite is true. When a road and a railway merely intersect each other in a level crossing, both symbols continue uninterrupted.

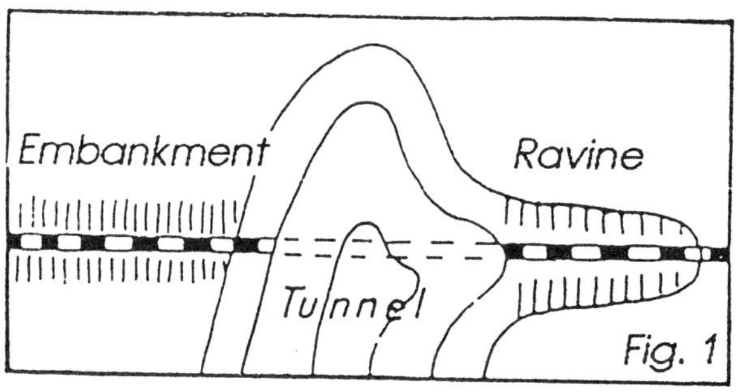

Fig. 1

Roads, railways, rivers, etc., are not marked in their natural width, but wider; remember that 1 millimeter on the map in a scale of 1:100,000 corresponds to 100 meters on the ground.

In this handbook we use the symbols of the Reichskarte. All additional symbols and abbreviations are described in the

Appendix A at the end of the book.

The Map Texts

The type, position, and size of the map texts are adapted to fit the size and the shape of the terrain features on the map.

The only exception is the name of villages, towns, and cities, for which the name is always written from west to east.

The name of a watercourse is written from left to right and follows its course. The text may also go backward.

Forests and heaths are designated with vertical letters.

The names of high-altitude areas are followed by the actual altitude in parentheses.

The name of a town or village is followed by a number in parentheses, which gives the altitude of the town's church over the normal-zero level. The normal-zero level is based on the average sea level as determined by hourly readings over several years. Since 1912, a point near Hoppegarten in Berlin, marked by five pillars set in the ground, has ben designated as the base for all altitude measurements. This point is located 37 meters above the average sea level.

Terrain Symbols and the Shape of the Ground

The natural shape of the ground is depicted in the map in various ways. In Germany, the most common methods are

(1) the depiction of the ground shape through vertical mountain lines,

(2) the depiction of the ground shape through layer or strata lines (contours),

(3) a combination of these two methods.

1. The depiction of the ground shape through vertical mountain lines

This is, for instance, the method used in the 1:100,000 Reichskarte.

When using this method, the altitude levels are marked with narrow, vertical lines. A zero-degree increase in altitude (that is, flat land) is not marked by any lines and therefore remains quite white. An increase in altitude of 15 or more degrees will be marked and therefore appears dark. Increases between 0 degrees and 45 degrees are also characterized by the fact that the steeper a hillside, the darker its depiction appears on the map.

To recognize all shapes of the ground properly takes much practice and a constant familiarization with the terrain.

Fig. 2 shows how the individual terrain shapes will be depicted according to this method:

(1) peak

(2) saddle

(3) plateau

(4) mountain crest

(5) hollow

(6) ravine

When using a map of this type, remember that the mountain lines follow the line of descent of water. This means that, as water flows, the highest point of a hill can be easily determined.

The exact altitude above the sea level can, with this kind of map, be determined only by comparing the terrain symbols with the printed altitude numbers following certain points.

Even so, this is sufficient for most of our purposes.

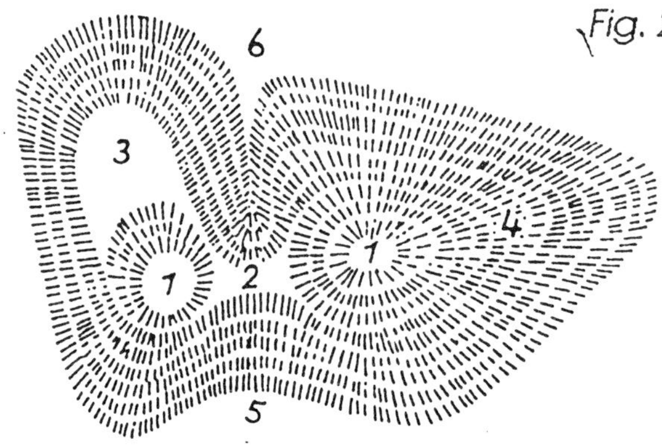

Fig. 2

2. The depiction of the ground shape through layer or strata lines (contours)

This method is, for instance, used in the 1:25,000 maps.

One can most easily imagine the appearance of the strata lines, as if one is slicing horizontal altitude sections of equal size (for instance every 10 meters). As each layer is positioned on top of another, we call these lines strata lines (Fig. 3).

The strata lines can also be called altitude lines or altitude curves, or contours. The reason is that points of equal altitude are connected by the same contour line.

It is easy to determine the altitude of any point of the map, as one merely needs to compare the contour lines. For instance, in the 1:25,000-scale maps of the Bavarian Topographical Bureau every 50-meter contour [is so] marked (for instance, 450, 500, or 550 meters above sea level). In addition, all 10-meter contours (for instance, 460, 470 meters above the sea) are marked with thin lines, all 5-meter contours (for instance, 455, 465, etc.) are marked with long lines, and all 1-meter contours are marked with short lines and the number of meters above the last 10 meters is included.

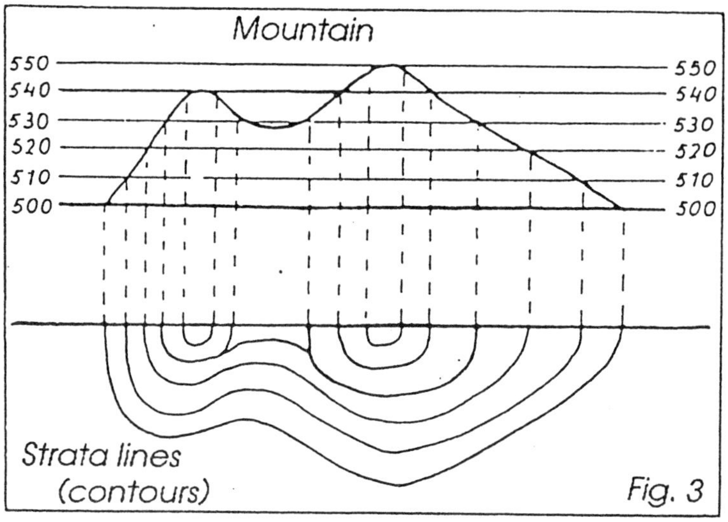

Fig. 3

In Prussia, the map charts in the scale of 1:25,000 indicate all 20-meter contours (for instance, 100, 120, and 140 meters), and all 10-meter contours (for instance, 90, 110, and 130 meters) as thin lines, all 5-meter contours (for instance, 95, 105, and 115 meters) are identified by long lines, and all necessary 1-meter contours are recognized by short lines.

A minor problem in reading the contours occurs when heights and hollows appear next to each other. Heights are distinguished from hollows in that the lower contours surround the higher. In a hollow, the opposite is true. Therefore, note that an exact and careful study of the contours is required. If no contour altitude numbers are marked on the map, then a small arrow indicates that the terrain feature is a hollow and not a height (Fig. 4).

In a contour map, the various types of terrain features are marked as follows (Fig. 4):

(1) peak
(2) saddle
(3) plateau
(4) mountain crest
(5) hollow
(6) ravine

3. The combination method of depicting both vertical mountain lines and contours

To achieve the advantages of both these types of maps, many maps combine both these features. So do, for instance, the mountain map charts of the 1:100,000 Reichskarte. In these map charts, the ascent indicators are not merely from 0 to 45 degrees, but from 0 to 80 degrees. Furthermore, to make the maps clearer, 100-meter contours are included.

Dividing the Map by Degrees

The Reichskarte is a so-called degree map, as the left and right side of the chart forms part of a geographical line of longitude while the upper and lower sides form part of a line of latitude. The exact latitude and longitude are noted on the side of the map. If, for instance, the upper left corner indicates 11° 20′ east of Greenwich, this means that the left map edge corresponds to 11° 20′ east of Greenwich (the meridian also used in Germany). The symbol for degree = °; for minute = ′.

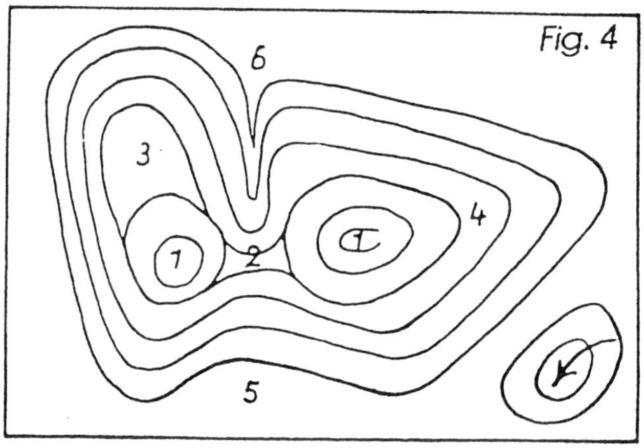

Fig. 4

The upper edge of each map always shows the north, the lower edge the south. The left edge is west and the right edge is east.

At the edge, the entire map is bordered with two lines. This border also indicates the lines of longitude and latitude in minutes. (Here the word minute does not indicate a measure of time, but a subdivision of the longitude and the latitude.) Any point on the map can be located in minutes of lines of longitude and latitude by simply connecting the minute markings on two borders at right angles with each other.

The Grid Net

All new map charts of the Reichskarte come with a dense grid net printed over the actual map. This grid does not correspond to longitude and latitude. It is well known that the distance between two longitudes becomes smaller the farther north one goes. The distance between the grid lines, however, always remains five centimeters. The purpose of the grid net is to facilitate the exact location of a given location, report this position without misunderstanding, and agree on the location of this given point.

The grid net consists of a net of parallel lines at right angles to each other and at a distance of 5 centimeters. The lines at the upper and lower map border (the x-axis) are marked "right-side numerals" (x numbers or x coordinates). At the left and the right side of the map border (the y axis), they are marked with the "high-side numerals" (y numbers or y coordinates).

A given point is therefore described first with the "right-side value" (read at the upper or lower map border) followed by the "high-side value" (read at the left or right map border). For instance, the location of Pasing in Munich is described as "right $^{44}60$ and high $^{53}35$" when using this method.

However, as there are always 5 centimeters between the lines of the grid net, we need a means to locate any point between these lines. For this purpose, we use the point locator.

The Point Locator

Every map of the Reichskarte is supplied with a point locator. The point locator is cut out and glued to a piece of strong cardboard.

With this instrument, we can easily and exactly determine the location of any point on the map.

It is very easy to use the point locator. You put the upper edge of the horizontal part of the point locator so that it lies close to one of the horizontal grid lines and so that the vertical part of the point locator is touching the point on the map whose location you want to determine (Fig. 5). You can now read the "right-side value" at the horizontal part of the point locator. First determine the next vertical grid line to the left, for instance $^{45}25$. You then read on the horizontal part of the point locator how far away the point you are looking for is from this vertical grid line. In this case, it is 2.2. The correct "right-side value" is therefore $^{45}25 + 2.2 = {}^{45}27.2$. (The first double-digit number, here 45, is mentioned to make certain that the users of this system have identical maps in their possession, something that obviously is necessary for the system to work.)

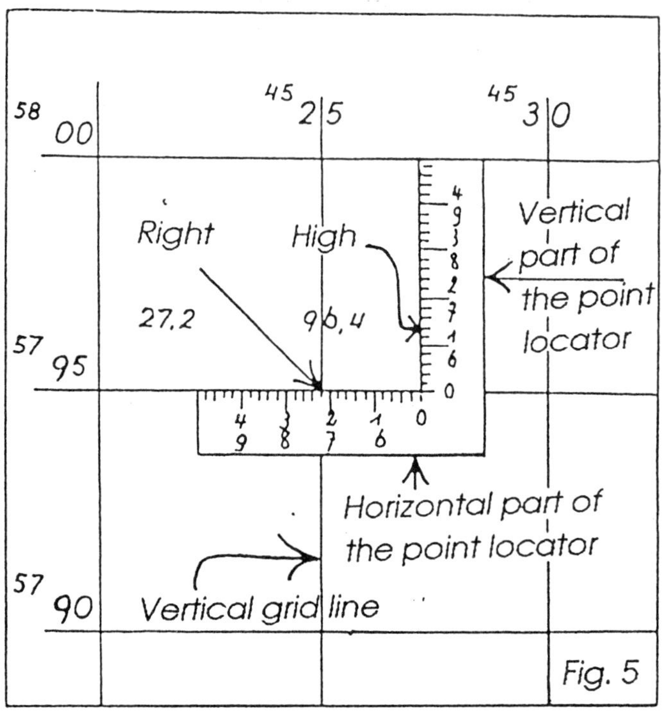

Fig. 5

The "high-side value" is then determined. Begin with the horizontal grid line, on which you have put the viewer, in this case $^{57}95$. Read the distance from this horizontal grid line on the vertical side of the viewer. In this case, the result is 1.4. In this way we get the "high-side value" of $^{57}95 + 1{,}4 = {}^{57}96.4$. In shortened form, we only write 96.4. (We have already checked that we are using identical maps.)

The correct identification of the given location is accordingly: Point A is located right 27.2 and high 96.4.

Needle Deviation and Magnetic Deviation

The term *needle deviation (from the grid line)* is not easily explained. Let us begin with the more common expression, magnetic deviation. The magnetic deviation is the angle in which every magnetic compass needle will deviate from true north. It is well known that the magnetic needle is not pointing at the geographic north, the North Pole, but instead the magnetic north, located on an island north of Canada in North America. The deviation angle is therefore the difference between magnetic north and true north.

With the introduction of the grid net, a common wish was to only use the grid lines, instead of the lines of longitude, when getting orienting on the map. This wish is now fulfilled. As the magnetic needle points to the magnetic north, and as one will use the grid lines as an aid for orientation, one must also consider the deviation of the magnetic needle with regard to the grid line. This needle deviation (from the grid line) also equals the angle between magnetic north and the direction of the north-south grid lines.

The magnitude of the needle deviation is nowadays indicated on the map. One must, however, consider that the needle deviation is constantly changing. Both the magnitude of change and the year of production of the map are printed on the 1:100,000 Reichskarte.

An example: From the map we learn that in Eichstätt [*translator's note*: the location of the Bavarian Police School] in 1926, the needle deviation was 6.45 degrees. As the annual decrease is 0.20 degree, the needle deviation for Eichstätt in 1936 is 6.45 degrees—(10 x 0.20 degree = 2.00 degrees) = 4.45 degrees. The needle deviation for Eichstätt is therefore 4.45 degrees.

This calculation tells us that to determine true north, we must align ourselves so that the position of the compass needle is not at "N" but at the point of needle deviation. The needle deviation must of course be determined for our current location. In Eichstätt, for instance, this point is between 4 and 5 degrees away from magnetic north. Only then will the "N" of the compass show true north (Fig. 6).

HOW TO FIND A TERRAIN LOCATION

If we are lost, we should not, as a rule, first turn to the map. Naturally, this does not mean that we have lost confidence in the map; to the contrary. But the map is to be used only when we reach a crossroads or some other difficult place. Then we should turn to the map, to briefly check that we are moving in the right direction.

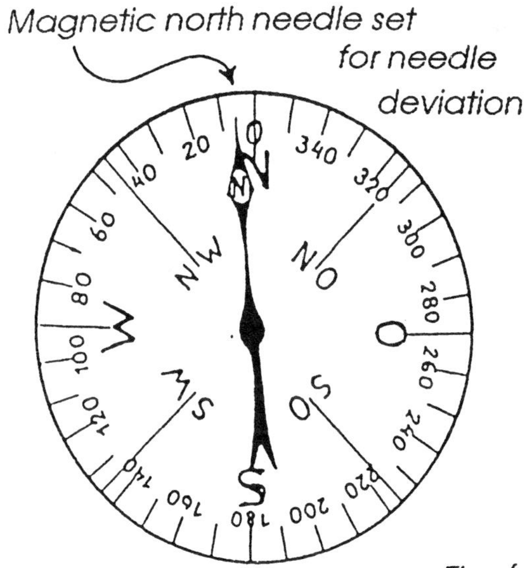

Fig. 6

How to Find a Location in the Terrain

Wherever we are, for purposes of orientation we should answer the following four questions:

(1) From where did we come?
(2) Where are we now?
(3) Where is north?
(4) What do we see around us?

Only when we treat our situation in such a systematic way can we reach our goal in the shortest possible time.

1. From where did we come?

Search on the map only for the last characteristic point that we have passed.

2. Where are we now?

The success of any mission depends fully on a correct and certain knowledge of our own position. If you are uncertain, compare the map carefully with the terrain. Also identify the hills and hollows marked on the map with those in the terrain and estimate the distance to at least two nearby already known locations.

We can also use imaginary lines in the terrain to reach our objective. On the map there may be two known locations, one to the right and the other to the left of our current position. Imagine a connecting line between these two points. By comparing the terrain with the map on either side of this imagined line, we can determine where we are. In doubtful situations, repeat this procedure with other known locations.

3. Where is north?

There are several means to determine north.

Right from the beginning, we must accustom ourselves to locating the cardinal points in the correct way and then confirm the obtained result by other means.

a) How to find north with the compass

Rough method:

Adjust the magnetic north needle of the compass with the correctly determined needle deviation. The direction, which is then marked by "N" on the compass (that is, not the magnetic north needle itself!) is north. Aim the upper border of the map in this direction.

Exact method:

(1) Position the compass on the map, so that the north-south line of the compass (not the magnetic needle) lies next to a north-south grid line on the map. (To accomplish this more easily, I recommend that you extend the north-south line of the compass with the help of a ruler.)

(2) The map and the compass are now rotated together, until the magnetic north needle is properly adjusted according to needle deviation. The map is then oriented toward the north.

For maps without a grid net, but with lines of longitude, or maps that have borders that coincide with the lines of longitude and latitude, north is located in a similar way. In these cases, the compass will not be positioned along a north-south grid line, but along a line of longitude or along the map border. Then one must take into account the real deviation, which is not strictly the same as the needle deviation. However, as the difference often is very small, for instance in Bavaria, we can still work with the needle deviation.

With loose map sheets, without the original border and without a grid net or longitude lines, we instead put a long ruler along the texts of the town names on the map (remember that the names of all towns are printed from west to east), and then position the west-east line of the compass along this line. Then rotate the map together with the compass until the north needle is adjusted according to the correctly determined needle deviation. This method works just as well.

b) How to find north according to the sun, knowing the approximate time

The sun stands at:
6 A.M. in the east
9 A.M. in the southeast
12 noon in the south
3 P.M. in the southwest
6 P.M. in the west
9 P.M. in the northwest

If you already know the direction of north, then this method can instead be used to determine the time without a watch or clock.

c) How to find north according to the sun with the help of a clock

It is possible to use a watch or clock as a compass. You hold the clock so that the short hand of the clock points in the direction of the sun. South is then in the direction, to be precise, of an imaginary line exactly between the short hand of the clock and the number 12 on the clock's face. In the morning this is calculated forward (as compared with the hour numbers), while in the afternoon this is done backward (Fig. 7).

d) How to find north at night

Navigation by night using the stars and the moon is not at all impossible.

The Pole Star always appears in the north. You find it by imagining a line extended from the two rear stars of the Great Bear constellation, about five times longer than the distance between these two stars (Fig. 8).

The Evening Star appears shortly after dusk in the west.

The Morning Star appears shortly before dawn in the east.

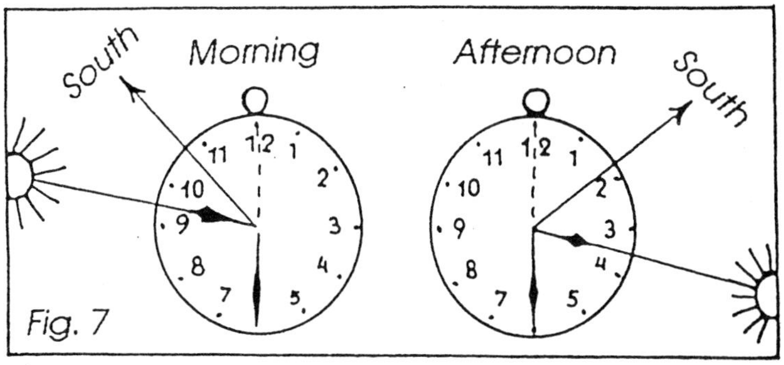

Fig. 7

e) How to find north according to the position of the moon

The full moon always stands exactly opposite the sun.
The (waxing) half-moon, first quarter, stands where the sun was six hours earlier.
The (waning) half-moon, last quarter, stands where the sun will be after six hours.
It then follows that:
The full moon stands at 6 p.m. in the east.
The (waxing) half-moon, first quarter, stands at 6 p.m. in the south.
The (waning) half-moon, last quarter, stands at midnight in the east.

f) How to find the general direction of north by other means

The west side of anything is usually the wind side. It is moss-covered and more weathered than the other sides.
The towers of old churches are usually at the west side, while the altar is in the east.
Vineyards are located on the south or southwest slopes of gentle hills.
Triangulation points usually have a cross marked with the four cardinal points.
And so on.

g) How to determine the four cardinal points with the help of the map

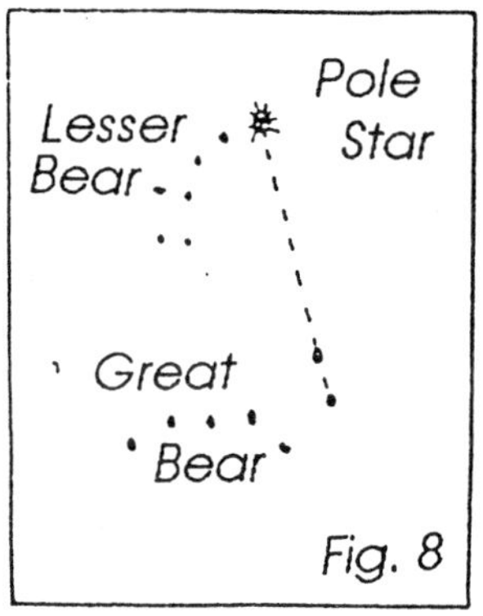

Fig. 8

First method: If you can see roads, rivers, or railway lines from your position, then rotate the map until the directions of these features on the map correspond with what you see in the real terrain. The top edge of the map now points to the north.

Second method: First locate a known, if possible a quite distant point that you can identify easily and correctly (for instance, a church tower). Then locate this point on the map. Connect this point with your own position on the map with a pencil or a ruler. Rotate the map until this connecting line on the map can be extended into an imagined line connecting with the distant point in the terrain. The upper edge of the map now points to the north.

4. What do we see around us?

(Terrain features to look out for and how to use these. How to actually locate one's position in the terrain.)

When we have determined our own location, and also the direction of north, then we can begin to find the way in the terrain.

First we must ensure that the upper (that is, the northern) edge of our map always points to the north. Accordingly, when we look toward the south, the map must be held so that north still points to the north. The only problem is to read the texts printed on the map. But you will soon get used to this.

When we hold the map in the correct way, we begin by locating the terrain features closest to us. Here, too, we must get accustomed to working systematically. We begin from the right and then turn to the left. To search first here, then there, only causes uncertainty and waste of time.

To read the map correctly without hesitation, we can use a ruler as a simple but easily available aid. I have already described one use of a ruler above in locating the cardinal points with the help of the map. Now we will look at yet some other possibilities.

If, for instance, I am uncertain about the correct identification in the terrain of Point A on the map, I will make a line on the map with the help of the ruler, connecting my own current position with the point in the terrain I presume to be Point A. Then I compare the terrain along this line with the terrain according to the map. If the map terrain symbols correspond with the actual terrain, then I have proven that the supposed point really is Point A. If the terrain symbols do not correspond, then I can determine whether the point I am looking for is located to the left or the right of the line, beyond or closer to the supposed but wrongly identified point. With a new line and a new comparison, I will soon reach my objective.

Another use of the ruler is to measure distance. Measurements on the map and comparison with the terrain will very often lead to correct results. When watching the terrain surrounding yourself, you must observe the following basic rules:

"At a distance, the terrain appears compressed."

"Closer, the terrain appears pulled out and wider."

Finding the Way in Terrain during Movement by Foot

To find the way during movement on foot in open terrain is not at all difficult. The only thing we have to do is to check the map in all difficult places, so that we follow the right way.

Difficulties only appear at night, in mist, or when moving through forested areas. To overcome these difficulties, we also have along with our proper map-reading skills the ideal aid: the compass. The standard-issue compass greatly assists in map reading.

However, even a simple compass is of great help, as it too can facilitate many tasks.

How to Use the Compass When Moving along Roads

It is common to find oneself standing, for instance, at a road crossing and wondering which road is the correct one. Here the compass will help us. The way to use it is as follows:

(1) Identify the correct direction of the roads on the map.

(2) Adjust the compass according to the deviation (needle deviation) and then determine the correct direction of the roads in the terrain.

An example: We are standing at a crossroads in a forest, and we have determined that the correct road runs in a southwest direction. The compass, correctly adjusted for deviation, shows us where the southwest is physically. The road that leads in this direction is the correct one.

However, for instance in large forests, new roads or paths are constantly cleared and old ones are abandoned, [so] we must get accustomed to using a control method:

At the beginning of the march, when we determine the travel (or compass) heading, we also divide the length of the march from point to point into paces (100 paces = 80 meters) or according to the time we need to cover this distance. If we get lost, despite the use of the compass, this will show us that we cannot be at the chosen point because the number of paces or the elapsed time is incorrect. Therefore:

(1) Determine the heading of the march (desired direction of travel)

(2) Determine the distance of the march

Cross-Country Travel—How to Set the Heading of March According to the Map

With a pencil, draw a line on the map from your present location to your objective. Then extend this line. Finally, determine the heading of the line. If it, for instance, runs north-northeast, then make a mental note: "Compass heading north-northeast."

(You can also, for instance, check the procedure by first aiming the map to the north and then determining the exact heading with the compass.)

How to Set the Heading of March over Terrain Where You Can Currently See the Objective of the March, But It Later Passes Out of View

(1) Adjust the compass according to the deviation.

(2) Determine the heading of the march objective with the compass. (It may help to lay a ruler next to the compass to fix the direction to the objective.) Read out the heading and, for instance, make the mental note: "Compass heading north-northeast."

When the compass heading is set, it is simple to follow the decided compass heading.

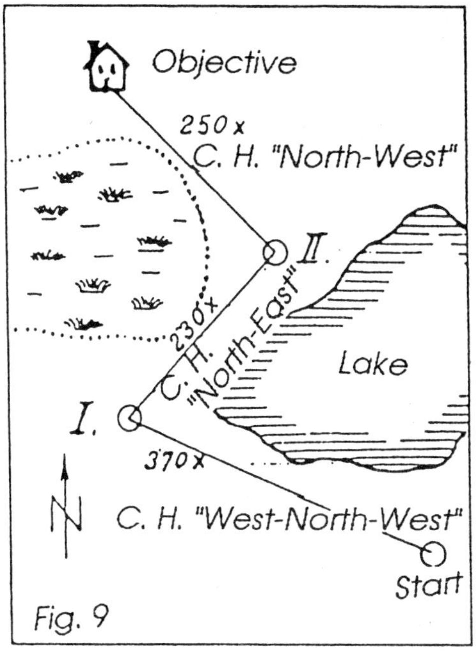

Fig. 9

How to Keep to a Straight Compass Heading

(1) Adjust the compass according to the needle deviation.

(2) Find the set compass heading (for instance, compass heading north-northeast) and march in this direction. At short, regular intervals, check the proper march heading.

How to March toward a Certain Point When a Straight Heading Cannot Be Used

(See Fig. 9.)

(1) Make an exact plan of the march on the map or on a separate piece of paper.

(2) Estimate the distance from point to point in paces (100 paces = 80 meters) or recalculated as elapsed time.

(3) Determine the compass heading (C. H.) from point to point.

Then follow this plan.

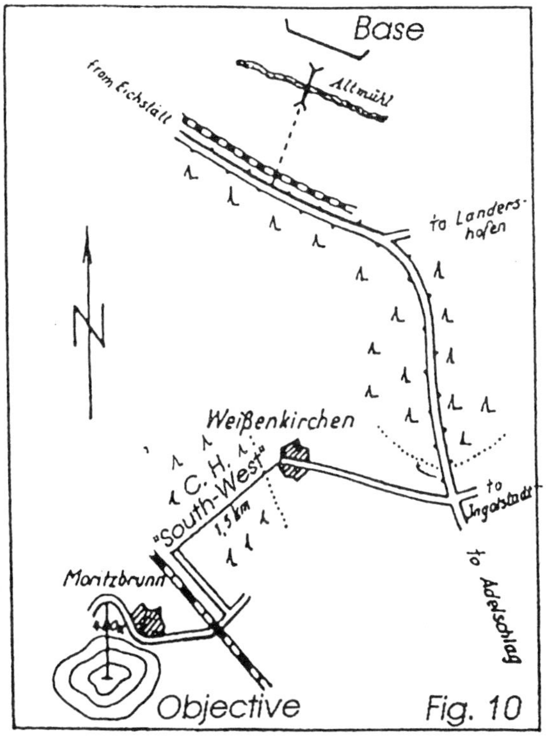

Fig. 10

How to Find the Way When Completely Lost

If because of carelessness you are completely lost, the following means can be used to determine your position.

(1) If possible, find an observation point from which you can see the terrain around you.

(2) Determine the location of north.

(3) Compare the way you have traveled with the last position known with certainty and the time that has elapsed since then.

(4) Compare the map with any significant terrain features, such as buildings, crossroads, watercourses, and woods.

Use the compass to determine the heading to these terrain features and compare their location to the map.

How to Find the Way without a Map

If we have no map detailing our route, we can still avoid going forth blindly, if we can first watch any map and from this prepare a plan of march (Fig. 10).

Such a plan of march must include the following:

(1) The names of places, forests, and so on, through which we will pass.

(2) Road forks and other roads, which we could mistake for the route of march.

(3) Whenever necessary, distances in kilometers, paces, or elapsed time from point to point.

(4) The direction of north and, when required, compass headings for our direction of travel.

You should be most careful whenever inquiring for distances and so on among the local inhabitants; you will often receive unreliable information, either because of ignorance or ill will.

Distance Estimation

Measuring distances is usually done with a pair of kilometer compasses or a ruler. But lacking such instruments, we can for this purpose also rely on simple means to make rough measurements. The length of a match is 5 centimeters, and the diameter of a 10-pfennig coin is 2 centimeters. Along with these, we can note the width of our little finger, the length of our thumb from nail to knuckle, and so on.

Time Estimation

In flat and reasonably hilly terrain one needs approximately one hour to cover 5 kilometers; for 1 kilometer, therefore, 12 minutes.

In mountainous terrain, you must also add approximately 15 minutes for each hundred meters of altitude difference. This added time can be decreased by very good roads and will be increased when moving across country.

HOW TO USE THE STANDARD-ISSUE COMPASS

Parts of the Compass

1. Compass bowl
2. Cover
3. Mirror
4. Ring
5. Sighting notch
6. Front sight
7. Lubber line ruler
8. Magnetic north needle
9. Index mark
10. Deviation
11. Bezel
12. Opening lever

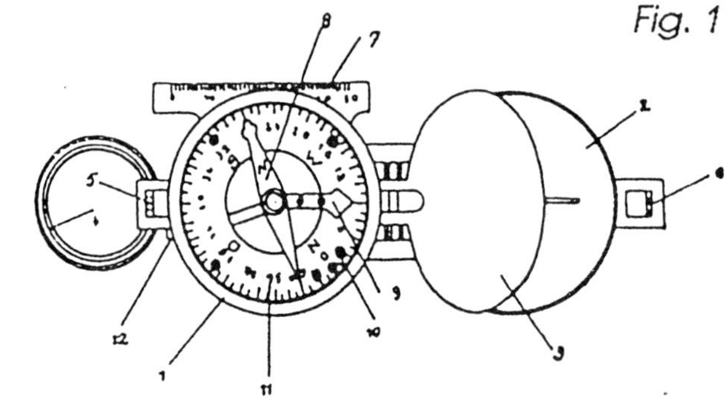

Fig. 11

General Information on Compass Use

When using the compass, objects made of steel and iron (for instance, weapons, high-power electric cables, steel helmets) must be kept away from the compass.

How to Aim the Map toward the North

a) Maps with grid lines

(1) Turn the bezel so that the index mark and "N" are aligned.

(2) Position the lubber line ruler along a north-south grid line, so that the index mark points to the north edge of the map.

(3) Rotate the map and the compass until "N" of the magnetic north needle and the deviation are aligned. The map is now oriented toward the north.

b) Maps without grid lines but with lines of longitude

In the same way as under (a), although the lubber line ruler is not aligned with a grid line, but with a line of longitude or the border of the map.

c) Maps without borders or irregularly connected map sheets without grid lines

(1) Align the index mark and "E" on the bezel by turning the bezel.

(2) Position the lubber line ruler along the text of a town name (printed from west to east), so that the index mark points to the east edge of the map.

(3) Rotate the map and the compass until the magnetic north needle and the deviation are aligned. The map is now oriented toward the north.

How to Set a March Heading When the Objective Can Be Seen but Later Passes Out of Sight

(1) Raise the compass mirror, and sight the objective over the sighting notch and front sight.

(2) Keep the sight line in your sight; turn the bezel with the free hand until the magnetic north needle is aligned with the deviation. (Can be seen through the mirror.)

(3) Read the numerical degree marking aligned with the index mark and make a note: "compass heading N."

How to Set a March Heading When the Objective Cannot Be Seen

(1) With a pencil mark a line on the map between your present location and the objective.

(2) Align the lubber line ruler of the compass with this line, so that the index mark is pointing toward the objective.

(3) Hold the compass level in this position and turn the bezel until the west-east line exactly follows the town name texts on the map.

(4) Read the numerical degree marking aligned with the index mark and make a note: "compass heading N."

Note: Naturally you may also first locate north and continue as we have described in a previous section, but in this case such a procedure is a waste of time!

How to Keep to a Straight Heading

(1) Align the compass with the already set heading. Raise the compass mirror. Align the magnetic north needle and the deviation, at the level of your eyes, through observation through the mirror. Retain the compass level in this position and sight over the sighting notch and front sight. Select one point visible along the sighting line as Point 1.

(2) March toward Point 1. There, in the same way, select Point 2, and so on.

How to Reach an Objective When the Direction of March Cannot Be Kept Straight

(See Fig. 9, above.)

(1) Prepare an exact plan of march on either the regular map or in a sketch map. (This procedure was described above.)

(2) In this plan, estimate the distance from point to point in paces (80 meters = 100 paces) or calculated as elapsed marching time.

(3) Determine a correct numerical compass heading for the continued direction of travel at each point.

(4) Follow the distance (in paces, etc.) calculated in the plan and the predetermined compass headings.

Note: This procedure corresponds fully to Fig. 9, except that here, for instance, "compass heading northeast" should be specified as, for instance, "compass heading 40."

How to Locate a Point Visible in the Terrain but Not Yet Located on the Map

(1) With the compass, sight the chosen point over the sighting notch and front sight. Align the magnetic north needle with the deviation.

(2) Put the compass, without rotating the compass bowl, level on the map. Rotate the entire compass—with the lower edge of the lubber line ruler at your present location—until the west-east line is aligned with the map texts.

(3) Draw a line along the lubber line ruler. The point you are looking for is somewhere along this line.

THE INTELLIGENCE REPORT (AND HOW TO WRITE IT)

The preparation of intelligence reports forms an important part of every police intelligence operation. An intelligence report is only of value when

(1) it is completely true and reliable,

(2) it is unambiguous and understandable

(3) it reaches the hand of the commander at the right time.

When do we need to make an intelligence report?

(1) Any first contact with the enemy.

(2) Any important change in the situation or status of the enemy.

(3) When any given street, town, forest, and so on, is confirmed to be free from enemies.

The time of the intelligence report is usually apparent according to the mission. Too many reports may be a disadvantage.

The orderly delivering an intelligence report should know the contents. The speed with which the report must be delivered is marked with crosses on the cover of the report. X = Not Urgent; XX = Urgent; XXX = Very Urgent.

The Head of the Intelligence Report

The following are the formal rules for all intelligence reports:

From Post: Here you mark the duty post and the immediate tactical duty position of the sender, not your name. Therefore, for instance, "Section A" or "Police Patrol Müller."

Report No.: Here you identify the number of reports that the post has already sent, for instance, "3rd Report." All reports must be numbered consecutively.

Location: The location must be clear and unambiguous. The location can be designated through approximately three to four indications, for instance, "the brickyard (= 1st indication) 500 meters (= 2nd indication) southeast (= 3rd indication) of Pietenfeld Church (= 4th indication)" or "Police Patrol 3/I Hindenburgstrasse 7."

Date: Write the month with Arabic, not Roman numerals, for instance 10.5.36 or 31.12.36.

Time: This is the time when the report was sent. The time is therefore not written until after the report is actually dispatched. Write the time according to the 24-hour system. Write the minutes in the upper position, for instance 10^{10} or 19^{30}.

To: This is the duty post to which the report is sent, not the name of the officer in charge. Do not write any honorific, such as *Herr* (Mr.) before the duty position. Write only, for instance, "To the commander of Section A."

The Text of the Intelligence Report

Always confirm whether you yourself have seen what you report, or whether the report is your assumption, or from whom you have learned the facts in the report.

Answer the following four questions:

When: Give the exact time of the observation. Also the hour!

Where: Give the exact location of where the enemy has been seen. If the enemy is marching, give the location of his column's front.

What: This includes everything that you have learned about the enemy, such as his strength, equipment, composition, organization, morale, the attitude of the local people toward the enemy, and so on.

How: Did I see the enemy, for instance, "deployed for combat" or "fortifying his area" or "resting, " or what?

What you, the sender, are doing: For instance, "The patrol continues along the northeast border of the interdicted area and clears any enemies there."

How to Sign the Intelligence Report

The reporting officer must write his name and rank below the report, to the right. Here he should write his name, not his duty post. Therefore, for instance, "Josef Huber, SS-Hauptsturmführer and captain of police."

General Information

The report must be written in clear, unambiguous writing and in German letters; only geographical and personal names may be written with the Latin alphabet. Ink and aniline may not be used. In an intelligence report text, there is no such thing as "to the right and to the left, on this side and on that side, in front and behind"; these civilian expressions may not be used. In an intelligence report, only the cardinal points may be referred to, for instance, "to the east" or "to the west." Street names should be written in the direction of march to the enemy; likewise, descriptions such as "entering town" and "exiting town" (with town names) should always be given in the meaning of the direction of march. Road forks should be written as "road fork Eichstätt-Rupertsbuch/Rebdorf" and road crossings as "road crossing Eichstätt-Rupertsbuch and Schernfeld-Pollenfeld." Expressions such as yesterday, today, tomorrow must be left out. Instead you should write the exact date.

Examples of Intelligence Report Texts

Sample report 1: The patrol observed at 14^{20} hours at the Eichstätt-Pietenfeld road, 200 meters north of the Eichstätt-Adelschlag and Pietenfeld-Weissenkirchen crossing, the front of an enemy column. Strength 300 men, almost all in the uniforms of the stormtroopers of the . . . political party. Armament: half of them with rifles, half with pistols. The enemy advances toward Eichstätt in military formation, with an advance guard.

The patrol is observing from the hill 1 kilometer northeast of Pietenfeld whether any more enemies are approaching.

[*Translator's note*: This sample report may refer to the SA of the Nazi Party, an armed political organization eventually suppressed by the Gestapo and the SS.]

Sample report 2: Based on information from a farmer from Pietenfeld, since 9:00 hours in Hellenbergerforst, 1 kilometer east of Adelschlag, loitering elements from the entire district are gathering. Strength approximately 100 men. No other information is known.

The patrol is moving along the west side of Pfünzforst toward Hellenbergerforst to ascertain the situation.

Sample report 3: We have determined that in Hindenburgstrasse, in front of the inn The Golden Ox, approximately 500 members of the . . . political party, wearing party uniforms, have gathered.

We will proceed there to determine the purpose of the meeting.

HOW TO PREPARE SKETCH MAPS FOR THE INTELLIGENCE REPORT

We distinguish between three types of sketch maps:

(1) Improvised sketch maps and situation maps.

(2) Sketches of the view from important observation points.

(3) Crime location sketch maps.

This chapter details the formal requirements to be adhered to when preparing any sketch map.

The Improvised Sketch Map or Situation Map

The situation sketch map should contain only necessary information and must be so clear that anybody will understand at once what the sender wants to tell.

The scale of the drawing can be any [desired], but the maker of the sketch must keep strictly to the scale he has chosen. The scale must therefore be noted at the upper edge of the sketch, and any (already written or printed) wrong scale must be crossed out. If the sketch map has no scale, then the sketch map must be marked "not to scale."

A sketch map must always be oriented to the north! If this is impossible, indicate north with an arrow!

Terrain symbols should follow [those of] the 1:100,000 map.

Explanations should, whenever necessary, be included in an empty space on the paper. These are only necessary, when additional map symbols other than the ordinary ones have been used.

For sketch maps included in intelligence reports, the following rule must be followed: All written names and comments must follow each other in the same style and direction.

Exceptions:

(a) The name of a watercourse, which follows its course.

(b) The name of a hill or mountain, which follows in a half-circle around the peak.

(c) The names of streets, roads, and railway lines, which follow their respective directions.

The Best Order to Follow When Preparing a Sketch Map

1. Decide what will be included in the sketch map.
2. Draw the necessary streets, roads, watercourses (in blue), and railways lines.
3. When drawing sketch maps of open terrain, include the towns, forests, etc. When drawing sketch maps of towns or cities, include necessary buildings and their corners, buildings in front, etc. Also include information on the number

of stories of the buildings.

4. Draw necessary hill contours (in brown or black).

5. Mark your present position with a blue X, and also mark other friendly forces (in blue). Mark the position of enemy forces (in red).

6. Add texts and any comments to the sketch map.

7. Add the title, scale, location of north, and your signature (name and rank).

The Sketch of the View from an Important Observation Point

When making observation sketches, an artistic treatment of the scenery must be avoided. The only terrain features that should be included are those that are important to the sketch. All terrain features must be depicted in a clear and simple style.

For this purpose, note the following: everything unimportant and all superfluous details must be discarded. Any hill must be drawn only in outline, to indicate its shape. Forests, trees, and hedges must be first drawn in outline, not exactly as they appear but according to their type. Deciduous forests, trees, and hedges should therefore be drawn as rounded shapes, while coniferous forests, trees, and hedges must be drawn in a zigzag line, as they appear in nature. Then fill in the outline of deciduous forests, trees, and hedges with slanting lines, and those of coniferous forests, trees, and hedges with vertical lines.

For trees without leaves, draw the trunk, a few branches, and a few thin offshoots.

For distant villages, groups of buildings, etc., draw only their outline.

When preparing an observation sketch of a building, in which enemies have fortified themselves, the purpose of a quick intelligence report is to include all important points, such as chimneys, roof entrances, windows, dormer windows, balconies, entrances, and so on. All these features should be numbered so that there is never any misunderstanding when referring to them. Also indicate the number of stories.

When the drawing is ready, first mark the friendly forces (in blue) and then the enemies (in red).

Then add any text comments, indicate the exact location of your own position, and finally write the title, direction of north, and your signature.

The Crime Location Sketch Map

The purpose of the crime location sketch map is to illustrate the exact situation in an area, where a criminal act has taken place, and the conditions surrounding this event. Such a map is then used in the investigation of this criminal act.

For a crime-location sketch map, exact measurements and a correct scale are required. If this is impossible, then at least two distance measurements must be added for each circumstance or object.

Apart from this, the same rules as apply for other types of sketch maps also apply here.

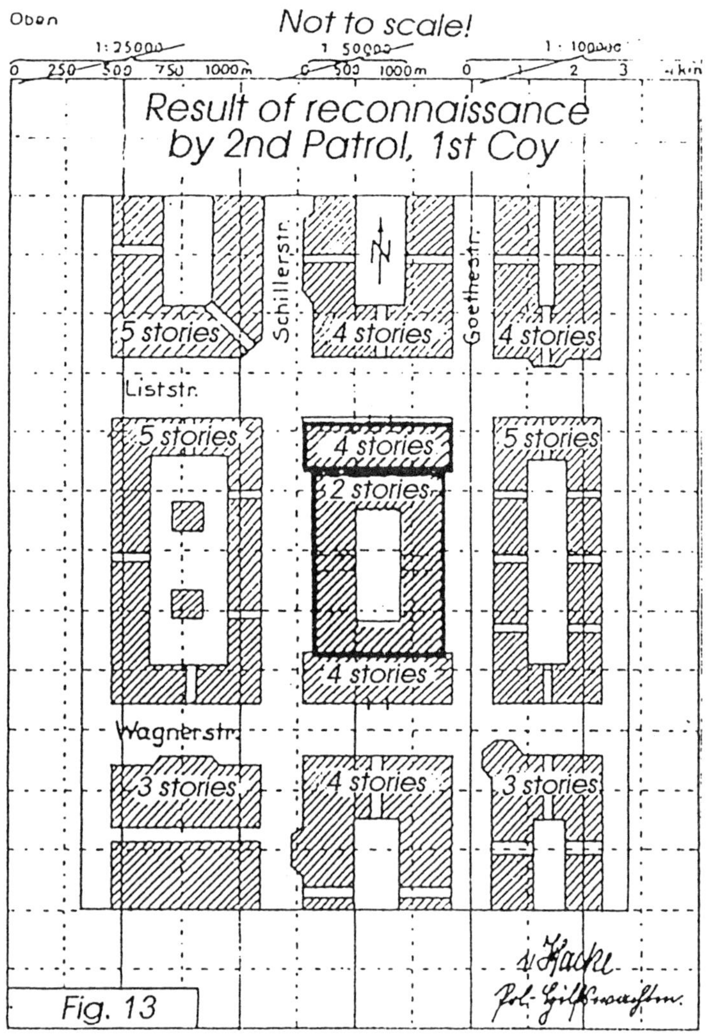

Fig. 13

AREA RECONNAISSANCE AND AREA ESTIMATION

We distinguish between police intelligence operations and police reconnaissance missions. A reconnaissance mission is always concerned with the determination of the local conditions of an area, whether it is open terrain or a city. An intelligence operation has the purpose of determining important facts about the enemy, his location, strength, etc.

The reconnaissance mission will survey the terrain to determine its use from a police tactical point of view. This information gives the commander an important foundation for his decisions and for the operation of the police force.

This chapter will explain how to plan, execute, and report a reconnaissance mission, and the value thereof.

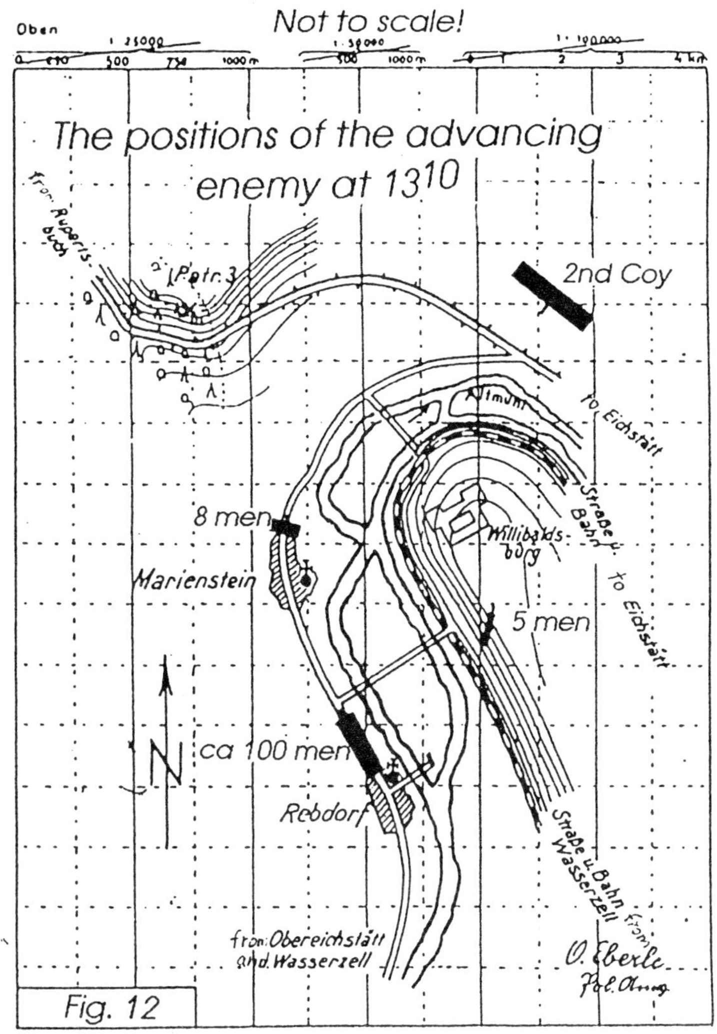

Fig. 12

The Reconnaissance Mission

The reconnaissance mission objective must be clear and well defined. The commander must give the reconnaissance team exact limits for the terrain to be reconnoitred. He must also indicate particularly which objects are to be looked for, as well as the police tactical purpose of the reconnaissance, as well as other pertinent information.

The reconnaissance mission is concerned with the following:

- The exact location of the objective, which may not be clear from the map, and its present condition
- The ease of movement in the terrain
- The best location from which friendly forces can secure the area and also control it through the use of covering fire
- The possibilities of finding cover from enemy observation and fire
- The suitability of the terrain for police tactical purposes
- The tactical conditions for the enemy
- And so on

How the Reconnaissance Patrol Executes Its Mission

The commander of a reconnaissance patrol will first repeat the mission order to the commander and then inform the other members of the patrol of the mission. By a careful study of the map, he will determine which objects are to be searched for, which route is most useful, and then plan how the mission can be executed in the time allotted for it. If there is sufficient time he will make a preliminary sketch map.

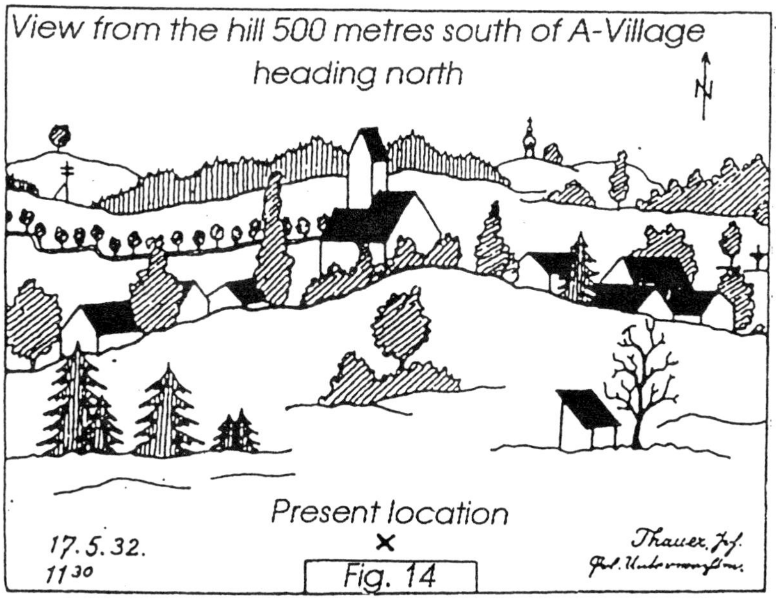

Fig. 14

"The reconnoiter must reconnoiter everything for himself. Because of this, he must rely strictly on himself, and himself only. To ask questions of local inhabitants is very risky. A steady comparison of the map and the terrain and the continuous making of notes and sketches of everything of importance are required."

Be prepared to destroy all notes in case of capture by the enemy.

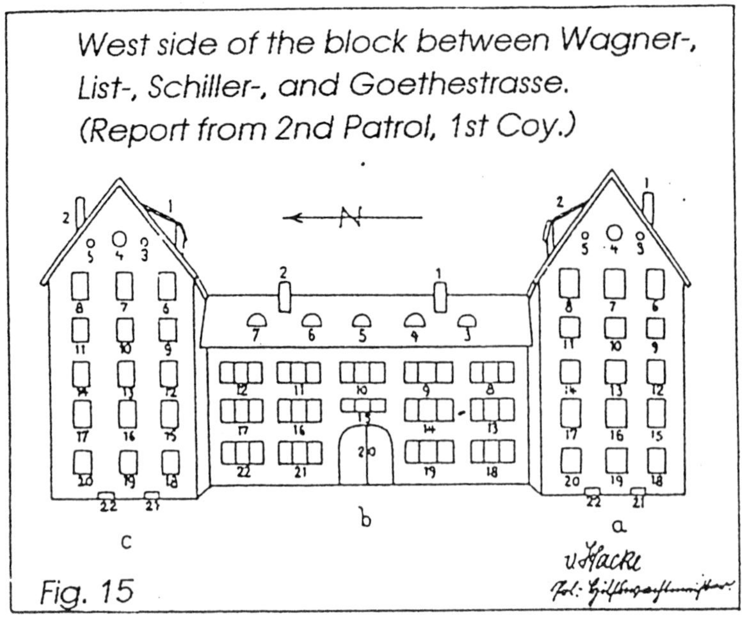

Fig. 15

Checklists for Fast and Correct Description and Evaluation of the Features and Conditions of Various Terrain Objects for Police Tactical Purposes

<ins>Streets and roads</ins>

—Ground conditions, paved or graveled, any eroded parts

—Impassable street sections, possibility to bypass such sections, repaired sections

—Street width (for horse-drawn vehicles 2 meters, for motor cars at least 2.5 meters, for two lanes 5 meters)

—Narrow street sections

—Bridges (conditions, load-carrying ability)

—Fords (depth for men on foot less than 1 meter, for horsemen 1.3 meters, current and type of riverbed also of importance)

—Ferries (steady and possible to use)

—Any changes in altitude—existence of trees

—Ditches

—Type of terrain beside the road or street

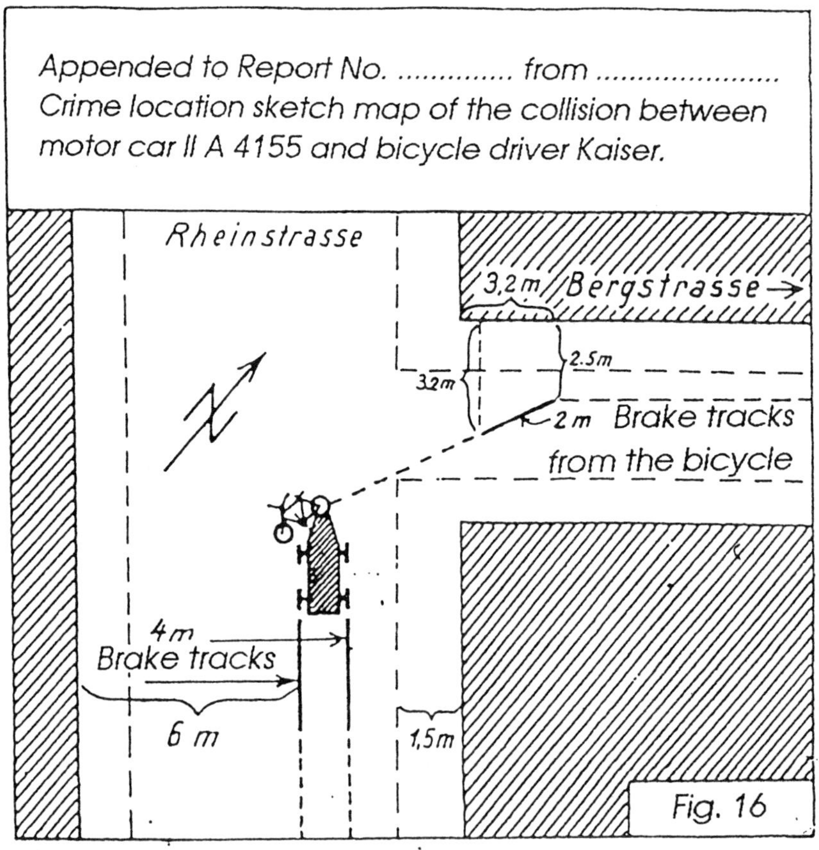

Fig. 16

<ins>Streets and public squares in towns</ins>

—The street's route

—Parallel streets

—Important buildings on the street

—Any courtyards in front of these buildings

—Entrances to the building open or locked

—Routes leading through buildings to other streets

—Where are semidetached houses, dormer windows, etc.?

—Which is the narrow side of a square?

—Are there any constructions on the square?

—Where do the side streets lead?

Railway lines
—Track gauge size
—Condition and location of station material (platforms, switch stand, and others)
—Condition of tracks
—Ascents
—Turns
—Embankments
—Ravines
—Bridges
—Tunnels
—Possibility of unloading and disembarking away from stations

Watercourses, rivers, canals, and marshes
—Length
—Direction of current
—Width
—Depth
—Banks and surrounding terrain
—Bridges
—Buildings
—Crossing possibilities (by barge, raft, ship)
—Suitable locations for this
—Fords

Ice thickness

A thickness of 10 centimeters is strong enough to carry a few people, also a horseman; a thickness of 15 centimeters supports a column whether on foot or mounted.

Woods and forested land
—General situation in the terrain
—Is the forested area controlling or being controlled by the surrounding terrain?
—Size
—Shape
—Forest limits
—Any altitude or area shape differences along the limits of the forest?
—Ditches or hedges in front of the forest
—The interior of the forest (brushwood, roads, areas where the trees have been felled, visibility, ground)
—The enemy situation

Towns and city neighborhoods
—General situation
—Is the town or neighborhood controlling or being controlled by the surrounding terrain?
—Size and extent
—Buildings close to each other or dispersed?
—The situation of the enemy
—Type and location of the streets
—Type of buildings
—Particularly important buildings

—Buildings from which one can control the surrounding terrain
—Telephone, telegraph, streetlights, and electrical lines

Single buildings
—Situation in neighboring buildings
—Garden
—Fences
—Number of stories
—Roof entrances
—Windows
—Basement windows
—Entrances
—Height of ground floor
—Which is the narrow side?
—From which neighboring building can one control all entrances and windows?
—Telephone wires and electrical mains
—Which surrounding terrain can be controlled from and viewed from the building?

The interior of a building
—Location of the rooms vis-a-vis each other
—Appearance and conditions of rooms
—Connecting passages
—Doors
—Stairs
—Windows
—Exits

For public buildings also:
—Cloakroom
—Toilets
—Small storage rooms

Hills
—Situation in the surrounding area
—Extent and shape
—Observational and field of fire conditions regarding the terrain in front, behind, and below
—Ease to climb

Roads in narrow valleys
—Extent with regard to length and width
—Limits
—Ease of passing through
—Possibility to close
—From where is the road controlled?

Checklists for Correct Description and Evaluation of the Characteristics and Suitability of the Terrain for Police Tactical Purposes

Reconnaissance for an assault or raid
(a) in open terrain:

—Terrain suitability for the approach and position of readiness of friendly forces
—Approach and access routes
—Where is enemy observation aimed?
—Where are good positions for preparing an assault or raid?
—Conditions of the actual terrain in the area of a planned assault or raid?
—Where are good fire positions, etc.?
—How is the terrain suited for a flank attack or envelopment of the enemy?
—Any covered access routes to the enemy position?
(b) in city areas:
Same as for open terrain.

Reconnaissance for defensive action
(a) in open terrain:
—Where are good defensive conditions, positions that the enemy cannot bypass?
—Where can access routes be closed?
—Where are good observation posts where one can see the main line of battle (and, especially, friendly defense zones)?
—Good defensive points
—Natural obstacles
—Clear areas allowing free fields of fire
—Possibilities to outflank the enemy
—For delaying engagements: covered waiting positions and access routes for the reserves
—Covered exit routes
For proper positioning of outposts, which are the best positions (covered terrain allowing access)?
(b) in city areas:
Similar to in open terrain: For outposts, read guard posts.

Reconnaissance for positioning an outpost
Location of the first perimeter: in daytime positions that allow good observation into the enemy-controlled terrain; at night at roads, that are easy to close.
Location for an outpost: in or near a position that must be defended from enemy attack.
Outpost reserves are, whenever possible, accommodated in houses.

Reconnaissance for accommodation
—Dry ground
—Protection against weather and wind
—Water supply and supply of firewood and straw within comfortable reach, therefore near an inhabited area
—Protection against enemy surprise attacks

The Reconnaissance Report

The best reconnaissance report is a simple but clear sketch map with short explanations. Ignore all nonessential matters.

POLICE INTELLIGENCE OPERATIONS

An intelligence operation is always concerned with the enemy. As long as one struggles against the enemy, the intelligence effort must never cease.

"Every patrol and intelligence section must receive a specific mission. The commander must clearly state what he wants to know. Every man in the patrol must know what the purpose of the mission is, so that the loss of the commander cannot prevent the success of the operation."

Intelligence operations can determine:

The location of the enemy, his approach route, his forces and their deployment, the position of his heavy weapons, fortifications and field works, extent of the area controlled by the enemy, his weak positions, general conditions, morale, armament, how the civilians regard the enemy, etc.

Execution of the mission:

—Repeat the mission order
—Familiarize the patrol with the mission objective
—Select a method
—Begin the operation
(Note: The duty of the intelligence organ is: Keep your eyes open!)
—Observe, before you are observed.
—Do not engage in a skirmish.
—Do not fire at the enemy.
—Circle enemy positions. Report whenever necessary.
—Proper use of binoculars can save travel time.
—Careful intelligence is worthless if is is not reported to the commander in the right time.

Reporting the result:

"Report orally, in writing, and through situation sketch maps and observation sketches. Signs and visual light signals agreed upon in advance often offers an imperfect but fast aid."

(Note: Reconnaissance and intelligence missions are seldom ordered separately from each other. As a rule, the same men will be sent on both intelligence and reconnaissance missions.)

HOW TO SECURE AN AREA OR A POSITION

The police in particular must expect that the enemy may strike from any direction. How to properly secure an area is therefore of the utmost importance. Too much security will exhaust the force; whereas too little can be of the greatest disadvantage.

The *objective* in every mission to secure an area is the protection of the main force against surprise attacks; crushing minor enemy attacks at the time of serious attacks give the police troops time to prepare and to prevent the enemy from observing our own situation.

For this reason, you must be strong, keep the available forces intact, observe everything before you are seen, as you must report to the security team what you have seen in the right time.

The strength of the security team depends either on numerical superiority, the availability of heavy weapons, or the existence of suitable terrain obstacles. In any case, it must be strong enough to observe the enemy, to report his actions, and to repulse any minor attack.

The security depends on

- the size of the security section,
- the extent of danger,
- the terrain (a river, for instance, can reduce the need for security forces),
- the visibility (at night and in fog, for instance, there is greater need for security forces).

The *deployment plan* of the security team must be chosen so that as large an area as possible can be observed; so that the security force will be protected from surprise attacks; and so that the enemy will be denied any insight into our own conditions. Good positions are hilltops, church towers, trees, and so on. Such deployment will save manpower, but you must make certain that the terrain that cannot be observed from one post can be observed from another post.

At night or in fog, security teams must be positioned on important roads and the terrain in between must be patrolled at irregular intervals.

The security team must not be farther from the headquarters than from where a report can reach the commander in time to make decisions. It is imperative that a report reach the commander in sufficient time for his force to prepare itself in case of a strong attack. The distance therefore depends onthe following:

- The enemy (whenever deployed nearby)
- The terrain (if visibility and observation ability are excellent, then a more distant location is possible)
- The visibility (the less visibility, the lesser distance will be acceptable)
- The tactical situation (when continuous defense has been ordered, then a deeper deployment—therefore more security posts, for instance, outposts—are required)

If the distance between the security team and the protected unit is too great, the enemy may bypass the security team and reach the protected area unchallenged.

PART II

•

PARAMILITARY POLICE OPERATIONS IN OPEN TERRAIN

FUNDAMENTAL RULES AND CONCEPTS OF POLICE OPERATIONS IN OPEN TERRAIN

This chapter will only treat lightly the specific duties of the police commander in each particular situation (these duties will be described in the next chapter). Instead, this chapter will give a short description of the fundamental rules and concepts of police tactics.

As the following text is concerned with paramilitary operations against partisans, insurgents, and rebellious rioters and dissidents, I have also included heavy machine guns in the descriptions and illustrations. Such support weapons are required when dealing with popular revolts.

Following the Versailles Treaty, the police force has only very few machine guns at its disposal. [*Translator's note*: This treaty, concluded after World War I, put limits on the number and type of armaments allowed to be exported to Germany. The terms of this treaty were later ignored, but neither the police nor the SS received heavy-support weapons until already well into World War II, in some cases only during the war against the Soviet Union.] In case of paramilitary operations against popular revolts, the Wehrmacht will supply the police with heavy weapons, especially machine guns. Support will also come from the troops of the SS. In this text, heavy machine guns are included to present a full view of how the police are to be deployed in paramilitary operations.

Paramilitary operations inside the city environment will be detailed in a later chapter.

Some Fundamental Concepts

Front, wings, flanks, rear.

Assaults and Raids

Every commander must always push his unit forward. Complete success can only be achieved through a correctly executed assault. The assault dictates the law to the enemy.

Types of Assaults and Raids

The objective of every police operation is restore calm and order. In a combat situation, the first rule is therefore to break the resistance of the enemy so that he is unable to later confront the police at another location; this is the only proper way to make certain that the enemy is brought to punishment. For this reason, it is always desirable to surround the enemy.

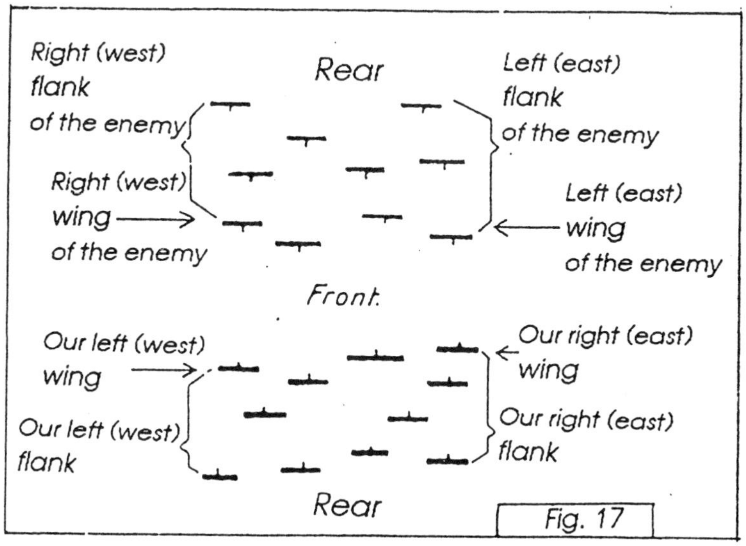

Fig. 17

How to Launch an Assault

When speed is important, the assault will take place directly from the deployment for movement, for instance, through deployment in dispersed formation to increase the combat readiness (Fig. 23), followed by deployment for combat. Most of the time, however, the assault will take place from a position of readiness. The position of readiness will be secured and protected through guard posts, submachine guns, and machine guns (Fig. 24).

The Assault Itself

The element of surprise is of utmost importance. If the enemy is surprised and uncertain about our direction and time of assault, then our operation is much facilitated. Therefore, maximum security must be taken in keeping orders and tactical movements secret to the enemy. Make feigned assaults or feigned tactical movements in other locations.

Intelligence and reconnaissance: The patrols sent out for these purposes can often also secure, along with [performing] their regular intelligence and reconnaissance duties, important terrain points in advance. Besides, these patrols also serve to hide the movements of our own forces.

Covering fire: When great haste is not required, always advance under cover of supporting fire. Covering fire can be from machine guns, submachine guns, sections, or individuals chosen by the commander. The purpose of the covering fire is to keep the enemy down when advancing. The best locations for covering fire teams are flanking positions or on higher ground.

Strike force: Every commander must always push his unit forward. A good moment to advance is when covering fire is available. Make use of every cover to prevent the enemy from seeing you. The advance follows, depending on the distance to the enemy, from dispersed deployment or deployment for combat.

A dispersed deployment or formation is used to increase the combat readiness of the force. The main unit is then divided into several smaller sections. The police company, for instance, is divided into police platoons, and the police platoon into police squads.

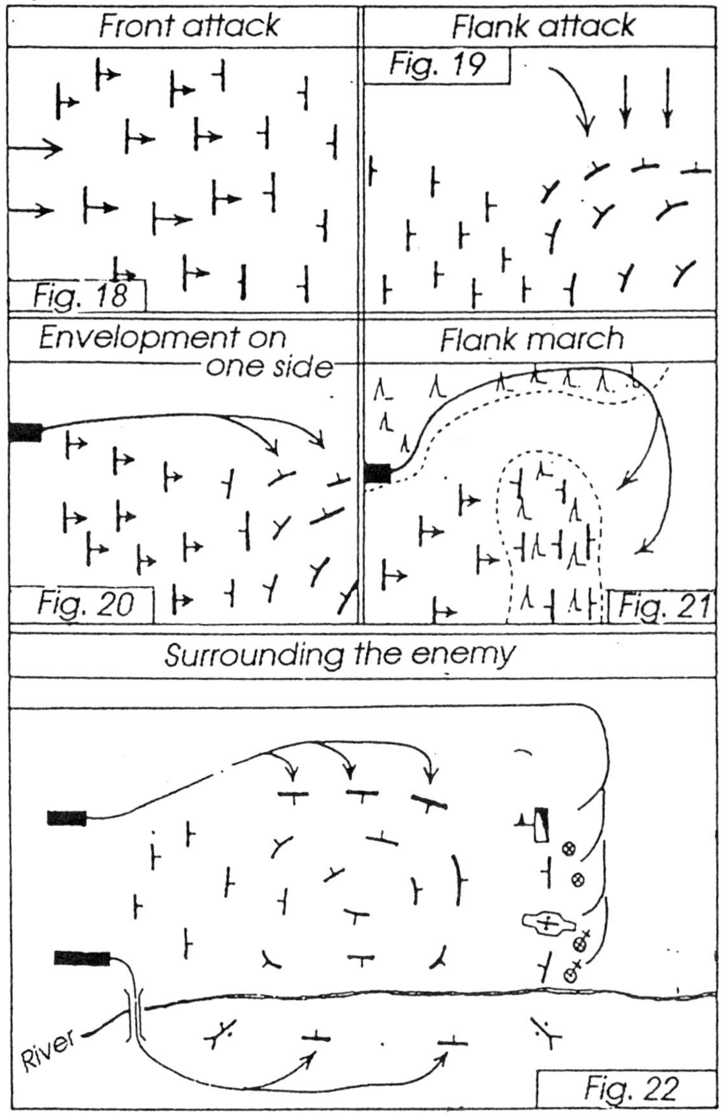

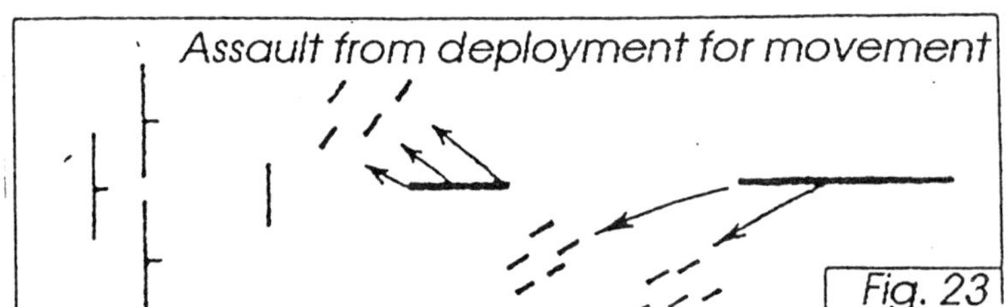

Deployment for combat is the formation of the police squads into columns and lines.

The column is used especially where narrow terrain cover is available, such as ditches, forest limits, hedges, road or railway embankments, and for flank marches.

The line is especially used because of its firepower effect at the front. It is also used in broad, covered terrain areas. The line is therefore the most common formation for skirmishes in open terrain.

The main point of assault depends on the situation, the mission, the deployment of the forces and weak points of the enemy, and the terrain (good approach routes, observation posts, and so on).

Breaking through the enemy positions: When one breaks through, the squads will immediately take control of the gained position and then pursue the fleeing enemy. Whenever the law so allows, maximum use of gunfire is authorized and should be used against the fleeing enemy. The squad will regroup and then continue breaking through into the depth of the enemy position, or prepare for defense against an enemy counterattack.

Pursuit

If the assault is successful and the enemy resistance is broken, then the pursuit must begin without any thought of the fatigue of the troops. "In tired troops, you will find the ultimate capacity."

Always strive to intercept a fleeing enemy. If possible, arrest many of the rebels, rioters, or partisans, but do not neglect to push after the enemy with all available forces. Especially remember to bring troops into his flanks and rear (interdict the terrain at bridges, rivers, and so on). Every pursuit must take advantage of any such possibilities.

Defense Operations

There are two types of defense operations: the continuous defense and the delaying operation. These should be treated in different ways.

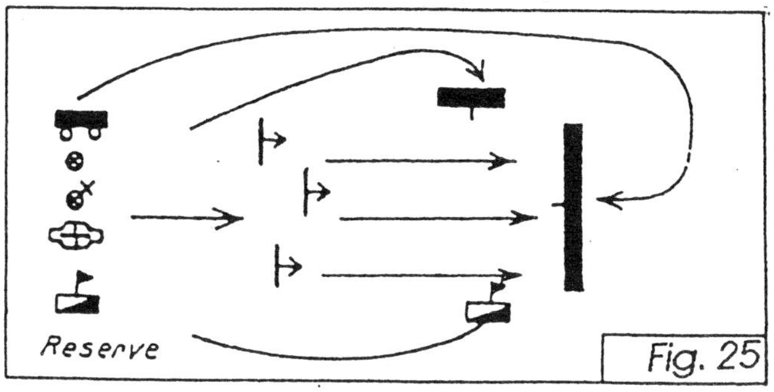

Fig. 25

Continuous Defense

Police troops defend themselves only against vastly numerically superior enemy forces, or to enable an assault in another location later. Then the police troops will attack.

Objective

The occupied position must be held under all circumstances. Repel any enemy attack. Prevent the enemy from breaking through your position. If the enemy breaks through, retake the position at once.

Combat method

Close defense to the last man. Retake any part of the position lost to the enemy through a counterattack.

The best deployment of defending police forces (no stereotyped pattern!)

Intelligence teams
Outposts
Troops deployed in deep formation along the main line of battle
Reserves

The Delaying Operation

It may be necessary, in order to gain time (for instance, when one does not wish to receive an attack from a numerically stronger enemy), for police troops to execute delaying operations against the enemy in certain locations. Such an operation, delaying the enemy by fighting a planned retrograde movement, can in good positions (at rivers, for instance) transform itself into continuous defense. A delaying operation can also—through short counterattacks with limited objectives—inflict severe casualties on an enemy who advances in a careless manner with dispersed forces.

Objective

Gain time and reduce the enemy strength while carefully preserving your own strength. Do not aim at any decisive conclusion, but retain your fighting strength.

Combat method

Switch between stubborn resistance and fast movement away from the enemy. Strive to keep the front wide. Limited counterattacks take advantage of favorable situations, when the mission so allows.

The best deployment of defending police forces (no stereotyped pattern!)

(See Fig. 26.)

- Intelligence teams
- Outposts—when sufficient forces are available
- Troops deployed in deep formation defensive zones
- Reserves

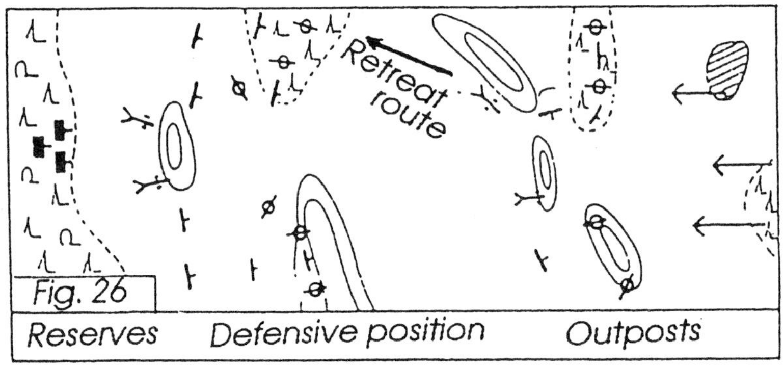

Fig. 26

Tasks of the individual components (explanation of the deployment above)

1) Intelligence teams

Planned combat leadership is facilitated through early and exact intelligence on the enemy (direction of movement, strength, composition, armament, morale, and condition of the enemy).

2) Outposts

These are the posts in front of the actual defense positions. Composition: machine guns, submachine guns, and carbine-armed policemen.

Purpose: to protect the police force from any surprise attacks; to give the troops time to prepare for combat; to make the operations of enemy intelligence teams more difficult; to mislead the enemy and thereby force him to deploy his forces too early and to split up his firepower.

Task: through deep deployment and movement reach the purpose above. To fight off enemy reconnaissance groups. Evade planned attack before the enemy comes too close. The retreat route must be planned so as not to disturb the field of fire of friendly forces.

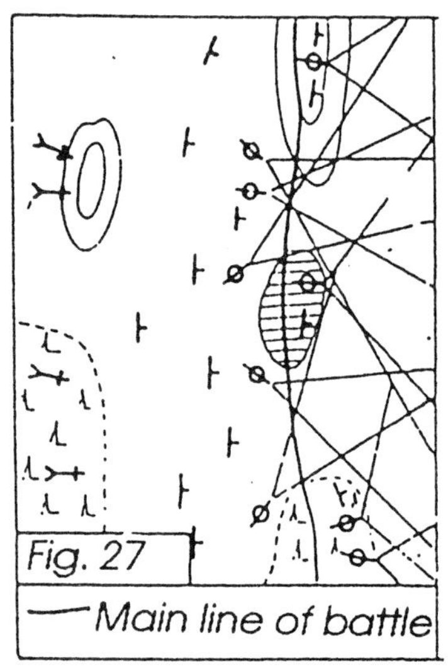

Fig. 27

3a) Troops deployed in deep formation along the main line of battle

Deployment position used in continuous defense. (See Fig. 27.)

The forward border of the defensive position, as a rule, forms the main line of battle. Position this line where you can concentrate the most powerful fire from all your available weapons at the attacking enemy and along a line that can be easily retaken if the enemy breaks through part of the position. At the end of the battle, the main line of battle must still be in the hands of our troops. Their situation depends most of all on the care with which the use of firepower has been planned. The terrain must be suitable, often behind a hill crest or through villages and forests; only in this way can the enemy be prevented from observing our positions.

Every troop must deploy both in depth and width through its terrain sector. Fortified positions, dummy field fortifications, and obstacles should be prepared. You must beware an enemy flank march. Prepare a free field of fire and locate any dead zones where you cannot fire at the enemy. Prepare communications, ammunition supply, and treatment facilities for the wounded.

3b) Troops deployed in deep-formation defense zones
Deployment position used in a delaying operation. (See Fig. 28.)

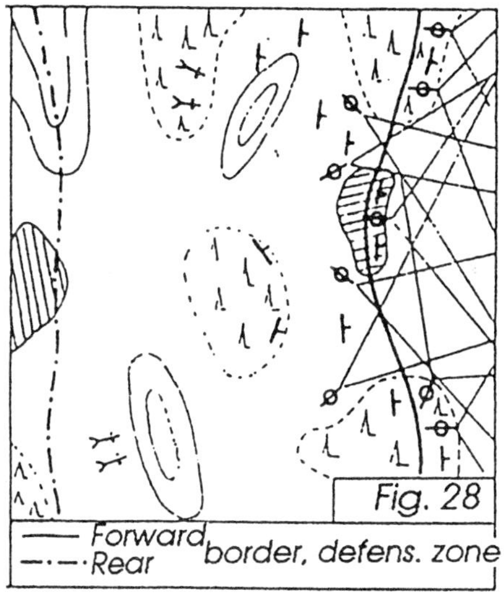

The situation in the defensive zone depends primarily on the effect of the firepower of all available weapons and only secondly on how to evade the enemy after combat. No main line of battle is fixed; instead, a forward border of every defense zone is chosen. Inside the defense zone, the battle is fought on the move. If the enemy comes so close to the forward deployed units of the defense zone that a prolonged defense of the zone in the long run becomes impossible, the zone may (upon orders from the commander) be abandoned. The troops will then withdraw under covering of fire and move into new positions at the rear of the defense zone. In most situations, however, the troops will, whenever possible, only withdraw at night. Deep deployment, fortified positions, etc., as in continuous defense, but with wider and deeper deployment.

4a) Reserves
Deployment position used in continuous defense.
Task: Be prepared to launch a counterattack or to intercept an enemy break-through.

4b) Reserves
Deployment position used in a delaying operation.
Task: To secure a planned battle and to secure the option to withdraw deployed police troops from combat.

Retreat

Retreat can only be authorized by the commander, and he must first make a thorough analysis of the tactical situation.

Retreat may occur for any of three reasons:

(1) Strong numerical superiority of the enemy compels small sections to withdraw from the enemy vicinity (the commander must make a thorough analysis of the situation before this is allowed).

(2) The commander can choose to pull troops out of one position, so that forces are freed for a decisive step at another location.

(3) The enemy has reached a position behind our unit, and a retreat will enable other police units to decisively attack the enemy.

For (1) and (2):

The best time to withdraw from the enemy without being noticed is usually after a successful armed action, in favorable terrain, or at night. Police armored cars, machine guns, submachine guns, and a few policemen remain in contact with the enemy. Through dynamic fire and switching between positions, they will trick the enemy into believing that the entire force remains in position. When necessary, they can even attack the enemy.

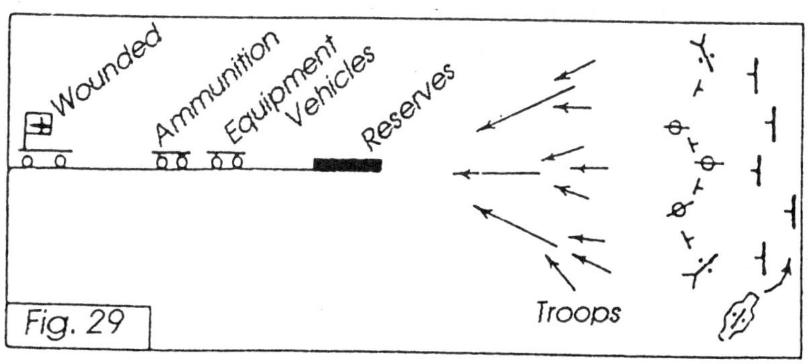

The actual procedure of retreat: transport of the wounded, heavy equipment and tools, ammunition, vehicles, reserves, and finally the troops, except those who remain in contact with the enemy. The latter troops will only be withdrawn when the main force has already taken new defensive positions, or when it is determined with certainty that the retreating main force has reached a satisfactorily safe distance.

For (3):

"If the enemy is behind, one must withdraw in small steps and connect the withdrawal with strong counterattacks."

Operations under Special Circumstances

Operations in villages and small towns

Here the police can meet two types of situations:

(a) The rebels, rioters, or partisans hold the town and avoid open terrain.

(b) The rebels, rioters, or partisans occupy the entire terrain sector including its towns.

For (a):

In this case, each town or village must be individually assaulted. For the operation inside the town environment itself, see Part III of this handbook, "Operating Procedures for Police Operations in the City Environment."

For (b), note:

"Town operations consume strong forces fast without influencing who will finally win." Therefore:

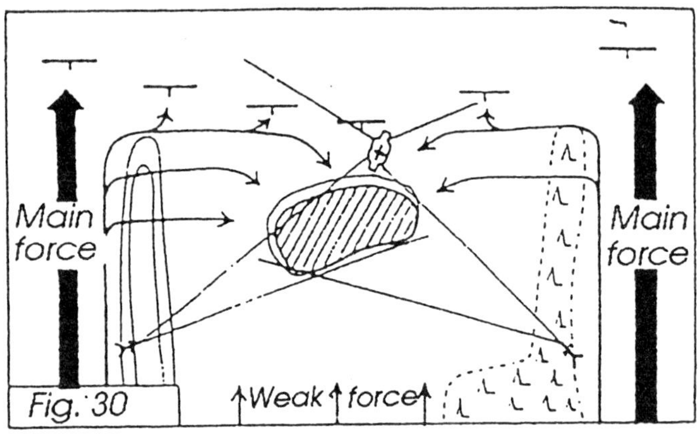

Fig. 30

a) Assault or raid on a village or small town (See Fig. 30.)

• The main force will bypass the town or village occupied by the enemy.

• Covering fire against the town or village will keep the enemy down.

• Units from the front of the main force will assault the town or village from the side and rear. Any available police armored cars are very suitable for this purpose.

b) Defense of a village or small town (See Fig. 31.)

• The main line of battle (MLB) will be positioned wherever is most suitable, in front of the village or town or drawn through it.

• Obstacles, roadblocks, flanking submachine guns, and machine guns permit us to engage the enemy at long distances and will bind his forces.

• Strong reserves will be positioned outside the town or village and under cover of the terrain. Their task is to fight off any enemy advancing toward the side or rear of the town or village.

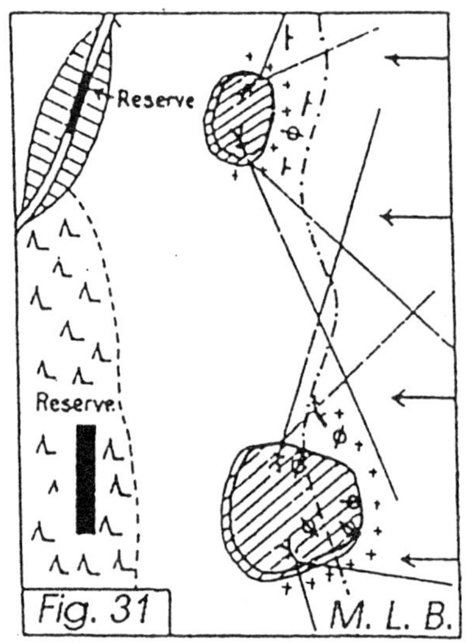

Fig. 31

Forested Terrain

The forest offers the attacker several difficult tasks (it is, for instance, more difficult to maintain a unified command—orientation and communications procedures are more difficult—and small arms fire has lesser effect); for the defender, a forest offers good possibilities of cover and defensive positions. Therefore:

Assault procedures

Go around smaller woods and take them from the flank or the rear (same as villages).

When dealing with larger forests possibly occupied by the enemy, advance under the protection of covering fire toward any part of the forest protruding from the limits of the forest (Fig. 32).

After taking this section, immediately regroup the troops.

For further advance into the forest, the best deployment is according to Fig. 33 (no stereotyped pattern!).

In this or a similar deployment, the troops will clear the entire forest sector by sector. After clearing one sector, always redeploy the troops. In smaller woods, the entire wood can be treated as one sector and the advance continued all the way through until you reach the forest limits beyond.

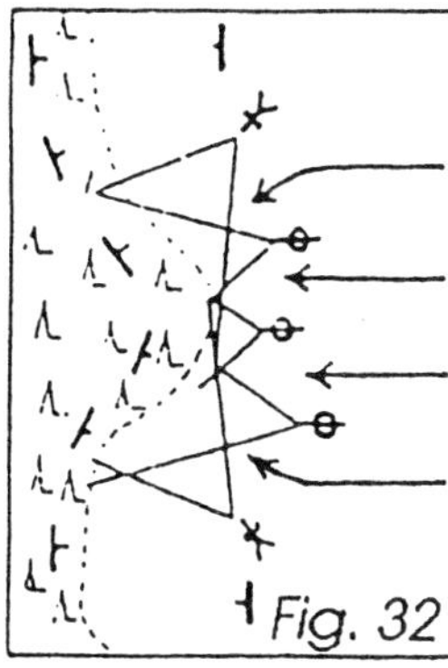

Fig. 32

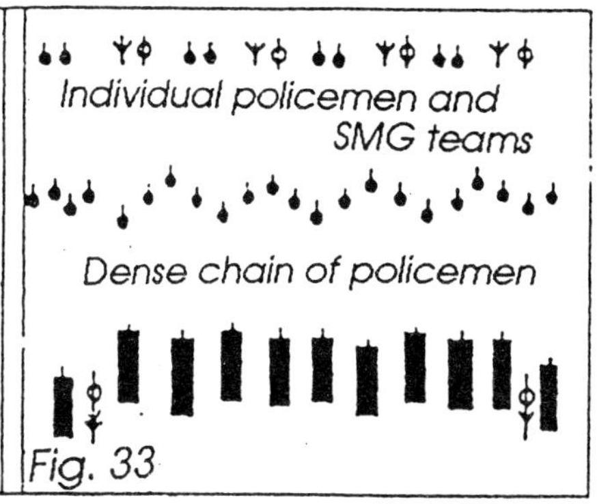

Fig. 33

Defense procedures

Position the main line of battle either in front of or through the forest. (The actual forest limits are too good and too recognizable a target for enemy machine guns, so this position should be avoided.)

Use individual trees, areas cleared of trees, and road crossings as positions for submachine guns and machine guns. Prepare many dummy positions!

If the enemy breaks through, throw him back by counterattacks or flank attacks.

Night Operations

Night operations demand well-trained troops; a few determined men can often turn a small operation into a complete success. The general rules for night operations are as follows:

- Secrecy regarding the purpose of the operation
- As simple and easy plan as possible
- Small distances in deployment
- Always maintaining and checking communications methods
- Detailed reconnaissance
- The greatest possible silence in execution

Assault procedures

Night assaults can have the following objectives: surprising the enemy, gaining suitable initial positions, or following up a previous success.

Disadvantage: It is very easy for the enemy to escape.

Defense procedures

Same as in daytime, although with due consideration of the general rules for night operations. If possible, use searchlights trained on the terrain in front of your positions, and always position reinforced observation posts. The reserves should be deployed near the battle line, but they should not be ordered into action until the situation is clear and there is no risk of surprise.

Operations in Low Visibility and Fog

The general rules for operations in low visibility and fog correspond to those for night operations. However, you must always expect that the fog may disappear at any time. Any order must therefore be promptly and decisively executed.

Operations in Ravines and Broken Ground

Narrow valleys and ravines primarily form obstacles for movement, so an exact reconnaissance of the terrain is very important.

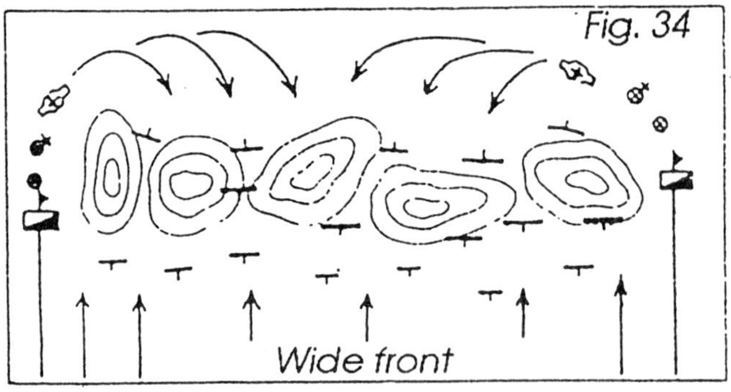

Assault

(See Fig. 34.)

Advance on a wide front. Every section will independently push on with all strength and speed, and if possible tie up many enemies at the front. Meanwhile, all mobile units will surround and encircle the enemy and take up positions on all sides around the enemy position. Assaulted from all directions, the enemy will be easily captured.

Pursuit

Aim to surround the enemy through fast envelopment of his position.

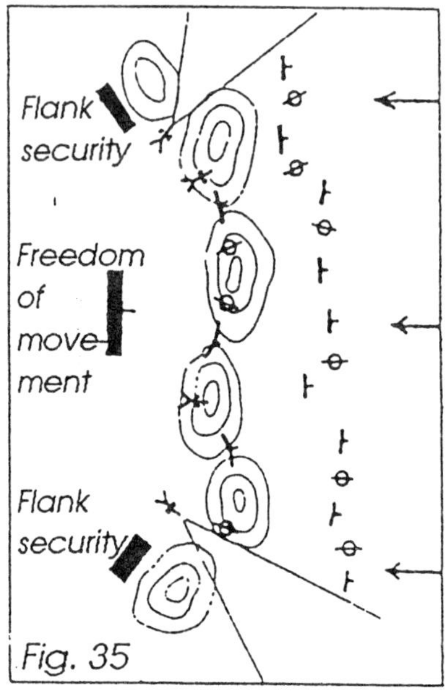

Defensive operations

(See Fig. 35.)

The main line of battle will be positioned either in part or in whole in front of the broken ground area on the side of the enemy. In case of only a few ravines or valleys, occupy both the hill and the valleys. To avoid an enemy flank march, position flank protective forces (flank security). Sections located behind the main line of battle and reserves should be deployed so that they can move in any direction.

River Crossings

The police do not have any pioneer (engineer) troops; we can only cross rivers by bridges or with the help of other means. We must therefore strive to take control of all bridges quickly and also to repair any destroyed bridges, so that they will be possible for men to cross on foot; this is the duty of each of us. We can only expect to receive assistance from the Wehrmacht pioneer units in times of all-out warfare. The following rules remain in effect even in such a case.

Assault procedures

Strive to get an insight into the situation in the enemy-controlled terrain by reconnaissance. Determine which bridges remain and which can be repaired, and prepare to take and repair these. Locate the best locations for crossing by other means, and secure the means for this.

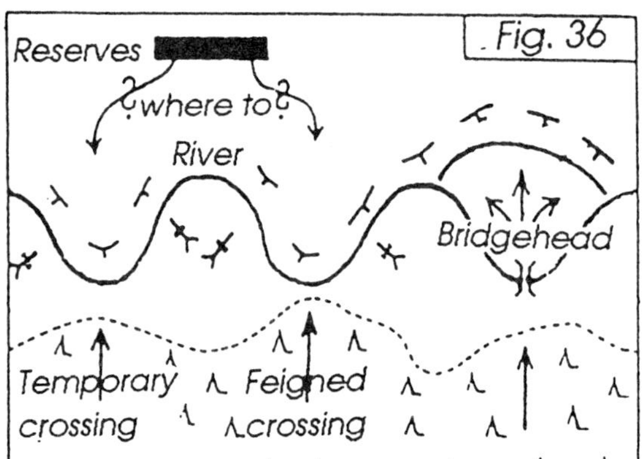

To cross by boats, etc., choose locations where the river bends against your direction of advance (Fig. 36). You will then gain the chance to unite the effects of fire and flank attack against the enemy on the other side of the river, even without actually crossing the river.

Advance under cover (if possible, at night), and use covered starting positions for the actual crossing. Prepare feigned crossings at several locations. The first actual crossed river sector forms a bridgehead. When the bridgehead is strong enough, continue the assault in the regular manner.

Defense procedures

Steady intelligence on the enemy (movements, position of readiness, purpose in crossing, and so on).

Keep advance positions along the river to prevent enemy reconnaissance and intelligence. Remove boats, barges, and other shipping means from the enemy side of the river and prepare bridges for destruction.

Prepare strong bridgeheads with fortified positions.

Keep the main force in readiness near the river. Prepare machine guns and submachine guns in positions where they can fire along the length of the river. At night, keep patrols along the river and light it up with searchlights.

OPERATING PROCEDURES FOR POLICE OPERATIONS IN OPEN TERRAIN

This chapter details actual operating procedures used in particular situations.

Maintaining Security During Movement On Foot

Customary deployment of police forces during a march toward the target of operations (no stereotyped pattern!)

(See Fig. 37.)

1. Plainclothes officers
2. Police armored cars
   Varying distance
3. Advance guard:
   Mounted advance-guard section
      Varying distance.
   Advance-guard section on foot
      200-500 meters distance
      Including orderlies.
   Orderlies and Communications Section
   Advance guard
      (Note: For smaller units, this section is not used [see Fig. 37].)
         300-600 meters distance
         Including orderlies.
   Main advance guard
         500-1,000 meters distance
         Including orderlies.
4. Main force:
   200-400 meters distance
   Including orderlies.
5. Rear guard:
   Deployment varying according to the situation and strength of the unit.

1) Plainclothes officers

Task: Provision of intelligence on enemy activities along and on both sides of the road or street, and in specific positions noted by the commander.

This task is solved by driving by bicycle or motorcycle in front of the column. Drive from point to point in the terrain. Conduct discreet inquiries among the civilian population.

2) Police armored cars (PAC:)

(See Fig. 37.)

Task: Security of the march. Provision of intelligence on enemy activities along and on both sides of the road or street and in specific positions mentioned by the commander.

At times also tactical duties—for instance, keeping a river crossing point open, interdicting enemy forces in a particular sector, occupying important points. Also distant reconnaissance of the terrain (conditions of streets, roads, bridges, negotiability of the terrain, and good observation points).

These tasks are solved by driving from point to point in front of the column. The distance to the mounted advance guard section will therefore vary.

3) Advance guard
(See Fig. 37.)

Task: The entire advance guard has the duty to guarantee the steadiness of the march, push weak enemy resistance to the side, protect the main force from surprise attacks, and, when encountering strong enemy forces, allow the main force time and space for deployment for combat.

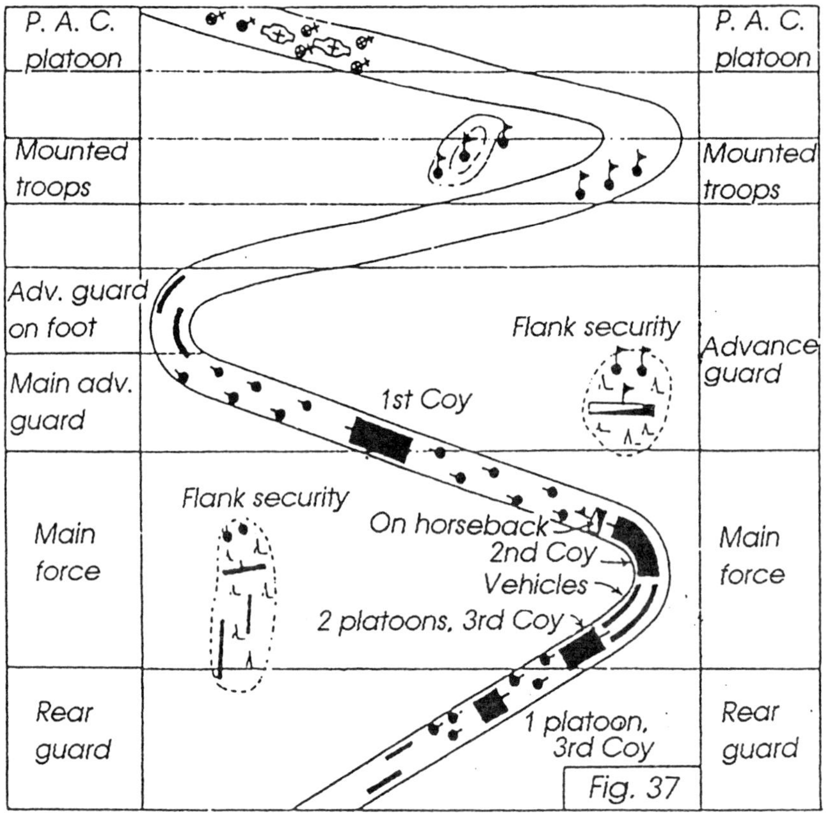

Fig. 37

a) Mounted advance-guard section

Task: Security of the march. Provision of intelligence along and on both sides of the street or road.

At the same time, the mounted advance-guard section must also fulfill the following tasks: Occupation (during the march) of important forward points along the route, such as railway junctions and bridges, careful search of larger forested areas, towns and villages, etc., which the main force must pass through, interdiction of narrow valleys, and the security of terrain sectors through which the main force must pass.

The mounted advance-guard section, too, solves its tasks by riding from point to point along the route. The distance to the advance-guard section on foot therefore varies.

b) Advance-guard section on foot

Task: Same as for the entire advance guard.

Strength and composition: One officer or senior police sergeant with 1–2 police squads.

Deployment: The advance-guard section on foot walks in column on both sides of the street or road, with two men sent approximately 100–200 meters forward as advance scouts, according to Fig. 38. (The commander of the advance-guard section on foot must prearrange signals for communications procedures with the advance scouts.)

Deployment and observation duties of the advance guard section on foot

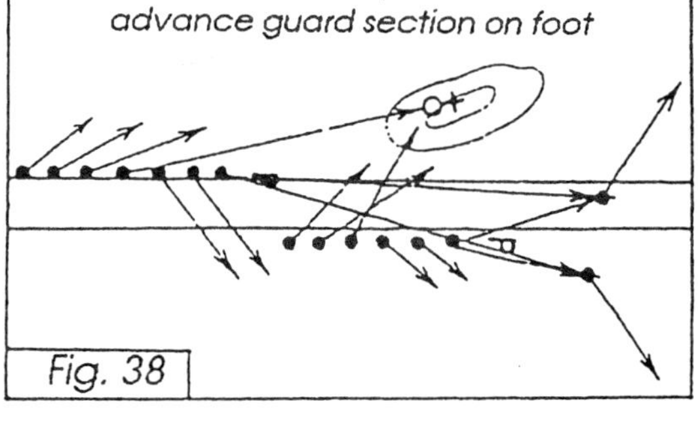
Fig. 38

c) Orderlies and communications section

Task: To maintain communications between the various sections. To ensure that the rear section follows the same route as the advance guard. To transmit messages and orders.

The orderlies are deployed in teams that split up when crossing terrain that cannot be well observed. In open terrain, the distance between the teams is approximately 50 meters.

At each road fork, the front man must inform the following police sergeant by sign which way to follow, and may only continue the march forward when the police sergeant has repeated the message. Short messages and orders will be transmitted from man to man. The man bringing the message or order must hear it repeated from the receiver, only then may he continue forward. Long messages or reports are transmitted from the first man to the next section in the most suitable manner.

d) Advance guard

Task: Same tasks as for the entire advance guard. If the unit is small (see Fig. 37), this section is not used.

Deployment: Closed march column.

The distance from the advance-guard section on foot is 200–500 meters.

The orderlies march between these sections.

e) Main advance guard

Task: Same tasks as for the entire advance guard. In addition, maintain flank security through the positioning of flank guard.

Deployment: Closed march column.

f) Flank guard

Task: Security of the march of the own column by following it along the flank and by searching any hills, woods, and similar terrain features at the flank of the marching force. The strength and composition of the flank guard depend both on the threat and the terrain. Either mounted police, men on bicycles, or a section on foot can be used as flank guard.

4) Main force

(See Fig. 37.)

Task: Closed, undisturbed march.

In case of contact with the enemy, the commander will retain the freedom of movement to make the right decision by maintaining a suitable distance to the advance guard.

The main force too must make use of a flank guard.

5) Rear guard

(See Fig. 37.)

Task: To secure the marching troops against disruptions and attacks from the rear.

The deployment of the rear guard depends on the situation and the strength of the total unit. The strength of the rear guard therefore varies from equal in strength to the advance guard to the strength of only a guard section.

Mounted police and submachine gun squads may be added to the rear guard, when such action seems required.

Maintaining Security during Movement by Car

During movement by car, the *deployment* of the police forces remains in the advance guard, main force, and rear guard in the same way as during movement by foot, described above. Only the distances between the different elements will be greater.

Intelligence is handled by plainclothes officers in fast vehicles, men on motorcycles, or police armored cars.

Task: Same as for police armored cars and mounted police, as described above. The vehicles in the same column keep only short distances from each other.

Communications procedure in the column is by sign from the members of the column. In addition, one man in each vehicle is selected to be responsible for maintaining visual contact with the vehicle in front and the one following.

Orderlies between the columns are generally not used during movement by car; for such purposes, approximately two motorcycle drivers are assigned to the commander of each column.

The *flank-guard* duties are handled by motorcycle drivers, police armored cars, or personnel in fast cars, depending on the situation. Their tasks are the same as during march on foot.

The commander will order a *degree of readiness.* When necessary, doors will be opened, back and side covers of the lorries will be removed, and submachine guns will be brought into position.

The personnel in each vehicle will be divided into *observation teams,* so that 1–2 men per vehicle will serve as observers at any time.

Every time a motor vehicle column stops, its personnel (every column independent of the others) will secure the area on all sides.

Maintaining Security during Movement by Railway

The best deployment of available
police forces (no stereotyped pattern!)
(See Fig. 39.)
(1) Forward security team
(2) Train guard
(3) Strike team
(4) Rear security team

1) Forward security team

Task: Protection of the train from disturbances from the front; and in particular protection of the train engine personnel. Observation toward the front.

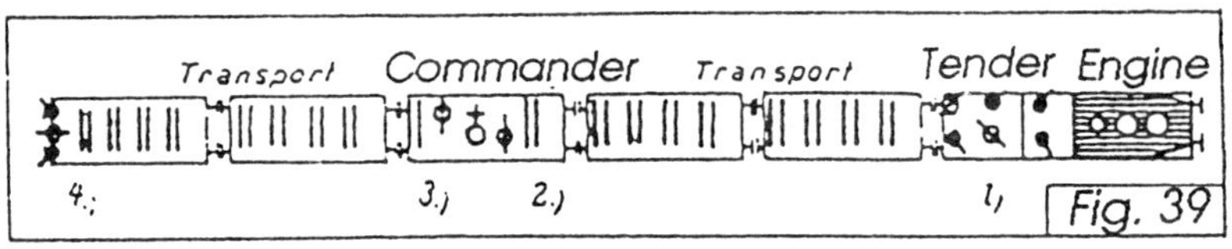

Fig. 39

Protect the train personnel and yourselves from sniping from the front and from the sides through defenses built from sandbags on the engine and the coal tender.

2) Train guard

Task: During movement, maintain observation on both sides. At every stop, secure the area around the train.

3) Strike team

Task: During movement, observation on both sides. At every stop, break any enemy resistance.

4) Rear security team

Task: Protection of the train against disturbances from the rear. Observation towards the rear. Maintain communications with the train commander.

Securing a Railway Station or Other Area during Disembarkation

(a) In an empty station or between stations:

The deployment and division (forward security team, train guard, strike team, and rear security team) of the guard detail of the train is abandoned when the train arrives at a station. Then the force immediately redeploys and prepares for disembarkation.

The forward security team secures the railway track towards the front.

The train guard and the strike team secure the train along the sides. Push away civilians who are hungry for news and take up positions around commanding positions, such as switch stands, station buildings, etc., to protect these by covering fire.

The rear security team secures the railway track towards the rear.

(b) In a station occupied by rioters or partisans and where disembarkation between stations is impossible:

Outside the station, the train will be stopped and a battle group without baggage will be disembarked. The battle group will take the station and secure the area both during the entry of the train and during disembarkation.

Checkpoints, Patrolling, and Guard Duty

Customary deployment of police troops (no stereotyped pattern!)
(1) Guard reserve
(2) Checkpoint police company
(3) Field guard details. These form the following:
- (a) Headquarters posts
- (b) Squad posts and double posts
- (c) Patrols within the post perimeter
- (d) Patrols against the enemy

Tasks of the guard post

Protect and secure the resting troops from surprise attacks, prevent the enemy from observing our positions, break minor enemy attacks, and in case of a serious attack give the troops time to prepare themselves.

A guard post must always be fully prepared against an attack. The personnel of the guard post must be ready for any sacrifice to protect the troops behind them.

1) Guard reserve
(See Fig. 40.)
Task: Reserves for the checkpoint guard detail.

To be located so that the reserves can have the best possible rest, but always remain ready for speedy deployment anywhere within the entire checkpoint sector, in case of sudden enemy activity in the secured area. The guard reserve is generally quartered in the town. To secure quarters, see under "Securing a Village or Small Town" (below).

2) Checkpoint police company (CP Co.)
(See Fig. 40.)
Task: Main security detail.

The commander will order appropriate intelligence and security measures; forward security perimeter (this is the line at which an enemy attack will be halted); responses in case of an enemy attack by day or by night; degree of readiness; construction of field fortifications; establishment of roadblocks; and deployment of machine guns.

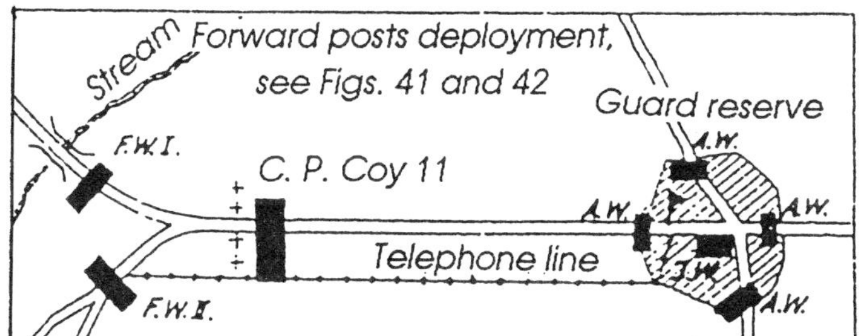

Fig. 40

The checkpoint police company is named after the company's number (for instance, Checkpoint police company 11, Munich Command).

3) Field guard details (FW)
(See Figs. 41 and 42.) Task: Security of the checkpoint police company and its guard posts. Continuous intelligence and observation.

Strength: From police platoon to police squad, depending on the task, the importance of the position, and the distance to the enemy.

Position: Important terrain points (bridges, railway lines, village and forest limits in the direction of the enemy, road junctions) with the best possible cleared fields of fire, under cover from enemy observation.

The field guard will often be assigned heavy machine guns.

Important field guard detachments are commanded by officers.

Field guard detachments are numbered in the police company in consecutive order from the right wing, with Roman numerals (for instance, Field guard detail II, Checkpoint police company 11, Munich Command).

a) Headquarters posts

Task: Immediate protection of the headquarters. Single sentries, positioned in front of the headquarters.

b) Squad posts (GrP) and double posts (DP)

(See Figs. 41 and 42.)

Task: Same as the tasks of the guard post.

Position: In daytime, so that the posts have a good observation point and, at the same time, can avoid enemy observation (camouflage, etc.). Trees, church towers, haylofts or haystacks, and buildings are particularly suitable. (Binoculars should be assigned whenever available.) By dusk, it is best to change the location of the posts, to avoid being caught unaware; good positions are streets and roads that lead from the direction of the enemy. By night, a denser chain of posts is often necessary. The sentries will stand in pairs. The individual members of one post observe together and must stand close enough that they easily can communicate and understand each other.

Double posts (DP) are relieved from the field guard detail and will generally not be more distant than 500 meters from this headquarters.

Squad posts (GrP) are positioned at important points (roads, commanding hills, etc.) and at more distant locations from the field guard detail headquarters (approximately 500–1,000 meters). Relief troops stay in the immediate vicinity. Important posts receive submachine guns or machine guns.

General post instruction: The commander will decide exactly how to stand or sit at the post location and how to behave. Usually the post members will dig protective foxholes and camouflage these. Smoking is allowed, unless the commander instructs otherwise. Nothing must be allowed to disturb the attention of the guard posts.

If the guard post observes something important about the enemy, one man will be sent to the commander with the appropriate message. If the post is attacked, it will signal this through several rapid shots in succession. Then a mobile patrol will be sent to assist and to exchange information with the guard post.

The guard post will allow each person, who is personally known, to enter or depart. Any other person must be interrogated and in uncertain cases be brought to the field guard detail headquarters or to one of the other guard posts (determined in advance by the field guard headquarters). Horsemen, bicycle and motorcycle riders, and cars must be challenged and halted; the latter must reduce their speed in the right time. If an order of the guard post is ignored, then—when use of weapons is authorized—the trespasser will be fired upon.

When it is dark, everybody who approaches the guard post—whose members will have weapons at the ready—must be challenged with a loud "Halt! Who is there?". If the trespasser does not stop following a third "Halt!", he—when use of weapons is authorized—will be shot.

A lone enemy leader with a small number of companions, who calls out that he comes to negotiate, as well as deserters from the enemy shall not be treated as enemies. They

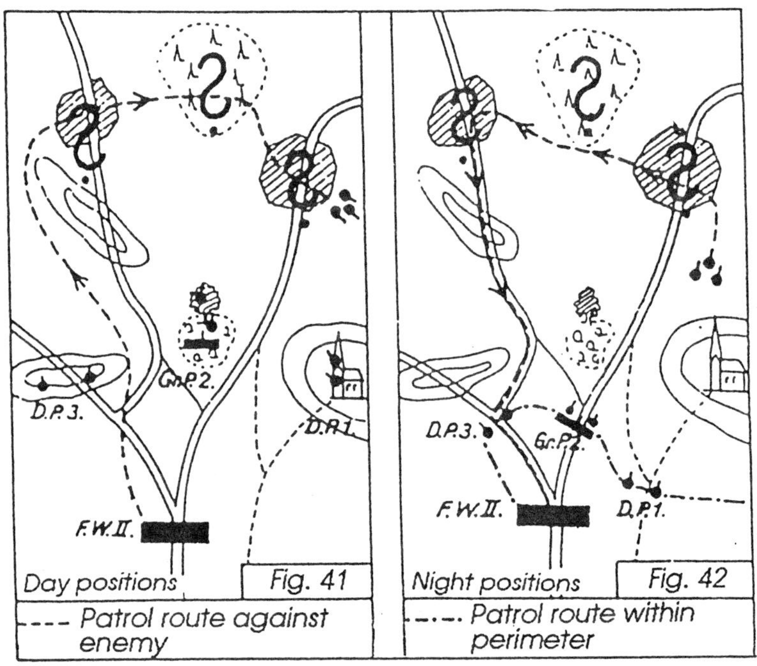

Day positions | Fig. 41
---- Patrol route against enemy

Night positions | Fig. 42
-·-·- Patrol route within perimeter

must lay down their weapons, and then—the emissary and his companions with eyes covered and without engaging in any kind of conversation—they will be brought to headquarters.

If possible, the guard posts will receive a sketch map of the terrain in front with the names of all villages and towns nearby.

The relief in the squad posts will be organized by the squad leader, while for double posts, the matter is handled by the headquarters commander. The commander of the relief troop must satisfy himself that the new posts know the general instructions, that the former posts impart any special knowledge of the situation or new instructions, and that the new posts correctly understand this.

Special post instruction:

(1) *Enemy*: Information on the enemy.

(2) *Behind us*: Location of the checkpoint police company, the field guard detail, and the roads to these locations.

(3) *To the right, left, and in front of us*: Location and identity number of neighboring guard posts. Communication procedures with these. Position of own sections positioned in the direction of the enemy.

(4) *We ourselves and our task*: Identity number of the guard post; orientation of the terrain; knowledge of any mobile patrols; and parts of the terrain to be particularly watched (visible road stretches, valleys, bridges that the enemy must cross to approach). Determine the most important distances in the terrain facing the guard post. Code words to be used.

(5) *How to behave in case of enemy attack*: Limitations in authorization to use weapons, if any.

Both double posts and squad posts are numbered from right to left, with Arabic numerals.

c) Mobile patrols within the post perimeter

(See Figs. 41 and 42.)

Task: To prevent the enemy from breaking through the post perimeter. To maintain the communications lines within the perimeter. Mobile patrol routes must be established to secure an unfriendly area. The number and strength of the patrols depend on the distance between the guard posts, the terrain, and the distance to the enemy. Each patrol consists of at least two men.

d) Mobile patrols against the enemy

(See Figs. 41 and 42.)

Task: To provide intelligence on the enemy and to familiarize oneself with the terrain, to provide information on this, and, in case of need, serve as guide through the terrain. (Keep your eyes open! It is better to observe than to be observed.)

While passing through the perimeter, the patrol must inform the nearest guard post of its mission and, upon returning, share its findings with the guard post.

General information

Any deployment of guard posts must be combined with sufficient protection from stationary or mobile patrols.

Securing a Village or Small Town

The best deployment of available police forces (no stereotyped pattern!)

(1) Town commandant and headquarters

(2) Resting troops (when necessary in alert quarters)

(3) Inner guard detail

(4) Outer guard detail

1) Town commandant

(See Fig. 43.)

Task: Supervision of combat readiness and security measures. Organization of domestic duties.

Cooperation and liaison with the Gendarmerie and/or other units from the ORPO (Public Order Police).

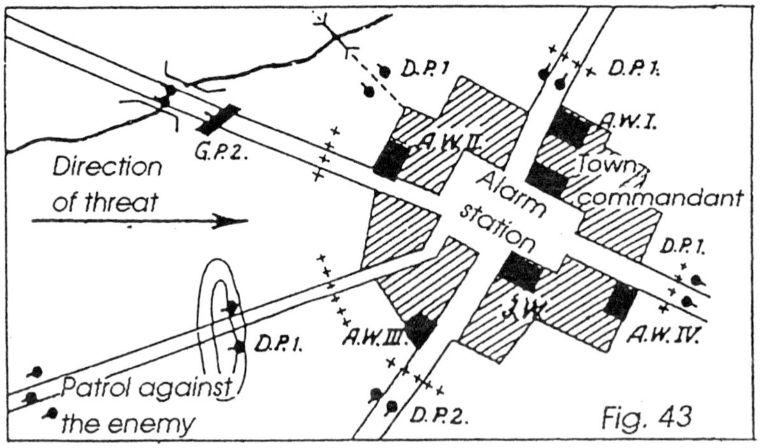

Fig. 43

2) Resting troops

Troops will be accommodated wholly or partly in the town, depending on the tactical situation. If not wholly accommodated in town houses, then part of the troops will be accommodated in nearby troop shelters on public squares, in gardens, courtyards, or outside the town. The commander will requisition quarters as needed.

There are several degrees of readiness, and the commander decides which one applies according to the tactical situation. If the enemy is nearby, readiness for an alert is required. Every man then has weapon and equipment ready, so that he can be put into action in the shortest possible time. Often it is necessary for each man to sleep in his uniform and for vehicle drivers to sleep by their vehicles.

The alarm station, the assembly area at which to gather when the alarm is sounded, must be well known to everybody.

3) Inner guard detail (IW)

(See Fig. 43.)

Task: To maintain security and order within the town in cooperation with the Gendarmerie and/or other units from the ORPO (Public Order Police).

The task is solved through patrols at irregular intervals. A sentry will protect the guard headquarters.

4) Outer guard detail (AW)

(See Fig. 43.)

Task: To secure the town itself. Prevention of any traffic in and out from the town. The outer guard detail is positioned at exits, roadblocks, and important points along the limits of the town and the surrounding terrain, so that it is impossible to surprise the resting troops. When necessary, the outer guard detail is assigned submachine guns and machine guns.

In the town, the outer guard details are numbered and identified by Roman numerals.

The outer guard details provide

(a) squad posts and double posts,
(b) patrols within the outer perimeter,
(c) patrols against the enemy,
(d) one sentry at headquarters.

The means to solve these tasks are the same as for the guard post (described above).

Securing Railways

<u>A. Securing railway stations and railway bridges and tunnels</u>

The best deployment of available police forces (no stereotyped pattern!)

(1) Guards at individual objects (such as bridges and tunnels).
(2) Patrols along the railway line.
(3) Mobile reserves.
(4) Informers for the organization of a report service.

1) Guards at individual objects (such as bridges and tunnels)
(See Figs. 44 and 45.)

Task: Ensure that the object is not taken by an enemy, who catches the guard unaware, or damaged by demolition work.

It is required that the guards are positioned for defense. Prepare and clear fields of fire and station guard posts. If necessary, construct barricades. Bring plenty of ammunition, water, and provisions (same as when defending a building).

Maintain communications with the commander of the railway security detail and the neighboring guard posts. (Besides the railway telephone, also prepare visual light signals or acoustic signals as backup warning systems.) Position guard posts at proper positions to receive signals.

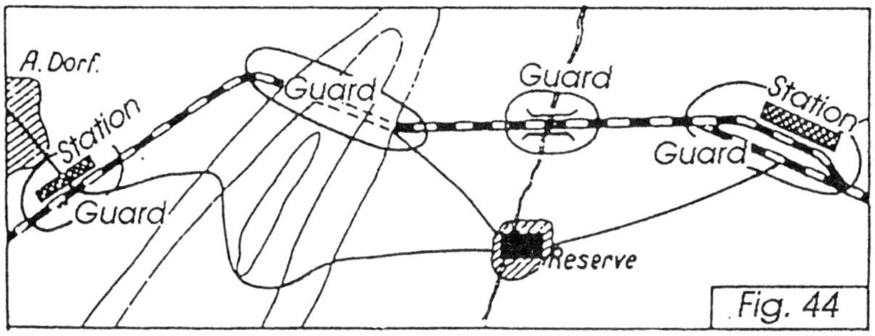

Fig. 44

2) Patrols along the railway line

Task: To maintain communications between the different guard posts. To search the railway line after disturbances. Remove any disturbances. Protect repair work and warn for approaching trains.

Best composition: 1–2 men railway police, 4 or more men regular police (this strength is necessary, so that 2 men safely can be sent as orderlies.)

Equipment: Along with the customary equipment, by day also two red flags, and by night two red lanterns, flashlights or similar means to warn for approaching trains; flare pistols are useful additions to the equipment.

Method of movement: Maintain as long a distance as possible but within line of sight, so that not all members of the patrol will fall in a surprise attack.

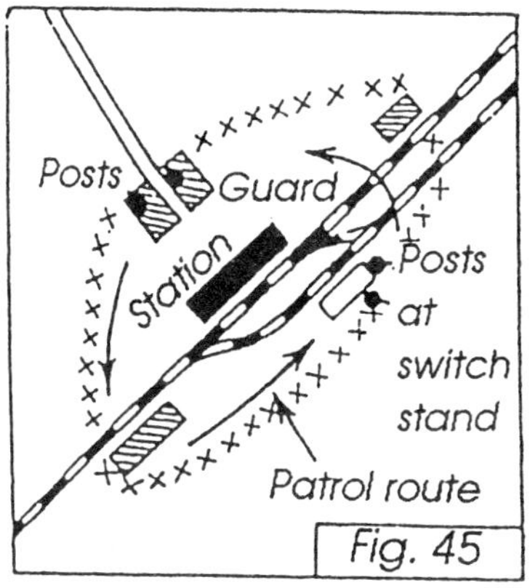

Fig. 45

3) Mobile reserves
(See Fig. 44.)

Reserves positioned at important and centrally located points. Motorized, mounted, or equipped with bicycles.

Task: To speedily come to the assistance of patrols or guard posts attacked by the enemy. Collection point for all reports on the enemy. To secure communications with the guard posts.

4) Informers for organizing a report service

Task: In cooperation with the Gendarmerie and through enlisting the help of elements loyal to the state, to organize a report service along both sides of the railway line. The informers will report all suspicious people, meetings, protest marches, or marches for other purposes, and so on.

B. Securing railway construction and repair areas

The best deployment of available police forces (no stereotyped pattern!)

(1) Guard posts around the workplace.
(2) When necessary: patrols used for patrolling nearby forests, towns, etc.
(3) Reserves at the workplace.

1) Guard posts around the workplace

Task: To secure an undisturbed progress at the work place. The guard posts must have a good field of fire, and when they are far from each other, each must be able to protect the flank of the other.

2) Patrols

If there are forests or towns nearby that the enemy could use as starting point for actions against the work place, then the mobile patrols must often search these.

3) Reserves at the workplace

Task. Be prepared for any kind of action.

C. Maintaining security on a railway train

For tasks, deployment and division of police forces in the train, see "Maintaining Security during Movement by Railway" above.

Methods of Raiding Secret and/or Fortified Headquarters of Political Opponents or Criminal Gangs

The timing of the operation, its preparations, and the role of surprise depend on the situation. Two cases are possible:

(a) We have sufficient time and are certain that the enemy cannot escape.

(b) It is urgent, and we must surprise the enemy at the objective.

The best deployment of available police forces (no stereotyped pattern!)

(1) Reconnaissance and intelligence by plainclothes officers (only in case a.).

(2) Interdiction units closing all access routes to the object from three sides.

(3) Raiding force.

(4) Reserves.

Raid and search procedure:

For (a): After a thorough reconnaissance and intelligence effort, the raiding force advances toward the target from starting positions far away from the objective, moving toward the common center, the objective, from separate directions.

For (b): In this case, the reconnaissance and intelligence effort by plainclothes officers is ignored. Instead the raiding force relies on the maps and information from the Gendarmerie for the exact location of the objective, and the force is deployed accordingly. The troops are transported in fast vehicles toward the objective. After disembarkation, each man rushes to his position, which is determined in advance, and the raid begins.

1) Reconnaissance and intelligence by plainclothes officers

Reconnaissance of the terrain conditions for the advance against the objective and for the interdiction of the area.

Intelligence on the location and strength of the enemy, etc. (report with sketch maps!), only in case (a); otherwise make use of existing plans and maps and information from the Gendarmerie.

2) Interdiction units

(See Fig. 46.)

Task: To prevent any disturbance from the outside, to prevent the escape of suspects, and to secure the area from civilian interference.

3) Raiding force

(See Fig. 46.)

Task: Arrest all suspects. Always advance with weapons ready to fire, and when required under cover of fire. Act energetically. Officers from the Gendarmerie will be added to the force.

4) Reserves

The position of reserves will be ordered according to the geographical situation.

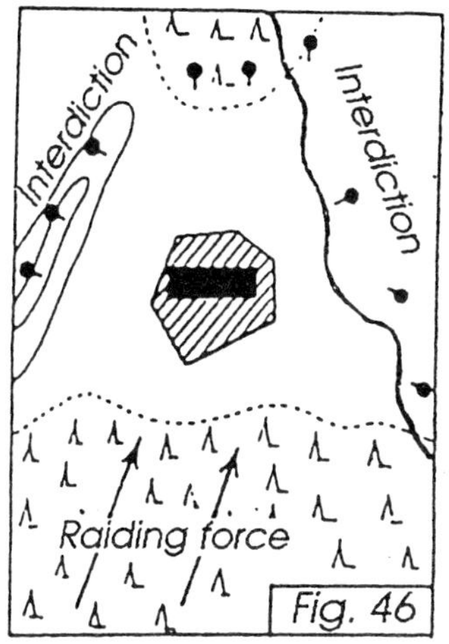

Fig. 46

Searching a Forested Area

<u>The best deployment of available police forces (no stereotyped pattern!)</u>

(1) Interdiction troops

(2) Search troops:

(a) scouts

(b) perimeter troops

(c) strike team

(d) reserves

1) Interdiction troops

Task: Maintain surprise when surrounding the forest to be searched, to deny the enemy any escape routes. The best composition of forces for this task: Mobile troops such as mounted police, men on bicycles, troops in motor vehicles.

2) Search troops

(See Fig. 47.)

Task: To search the forest. The best method is to begin the search from the narrow side of the forest. Search every tree from below to the top carefully and diligently—under maximum silence—and maintain continual visual contact with nearby troops. The strike team and the reserves will again patrol and search particularly dense parts of the forest, for instance, thickets.

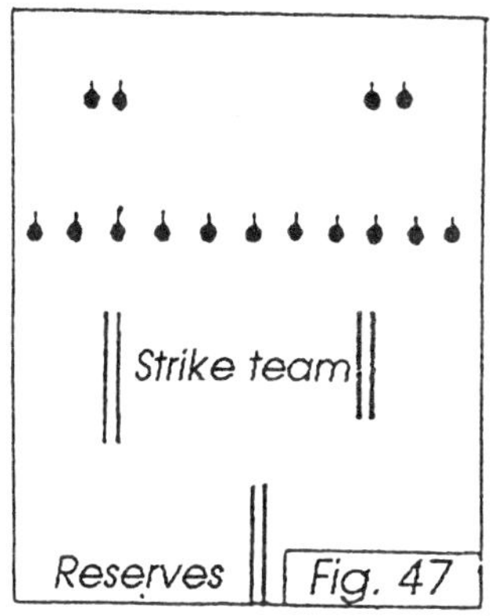

Fig. 47

PART III

•

OPERATING PROCEDURES FOR POLICE OPERATIONS IN THE CITY ENVIRONMENT

PEACETIME OPERATIONS AND OPERATIONS IN TIMES OF CIVIL COMPLIANCE

How to Block Streets and Public Squares in Times of Civil Compliance

Closing the street through a roadblock signboard (marked "Forbidden Entrance"), closing the sidewalk by guard posts
(See Fig. 48.)

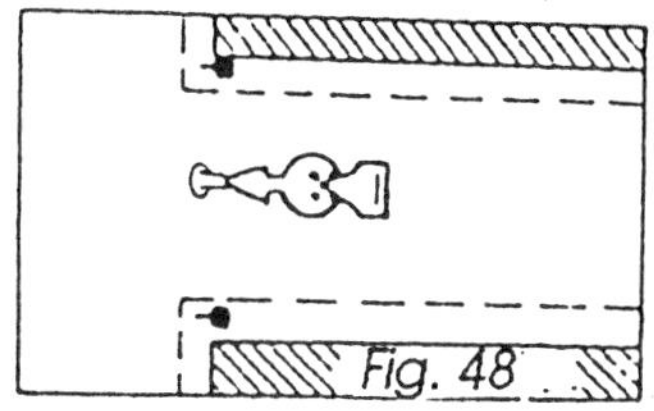
Fig. 48

The use of thick rope, steel cable, etc., to interdict a street when only a few policemen are available
(For instance in times of allowed demonstrations, in case of too few police officers, and so on.)
(See Figs. 49 and 50.)

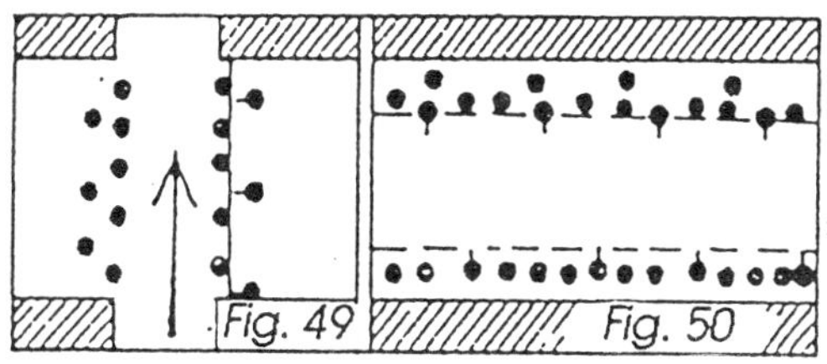
Fig. 49 Fig. 50

Closing the street by such police formations as the human chain, in one or two lines
(See Figs. 51, 52, and 53.)

Closing the street along the sidewalk
In case of few spectators: The front toward the demonstration (Fig. 54).
In case of many spectators: The front toward the crowds (Fig. 55).

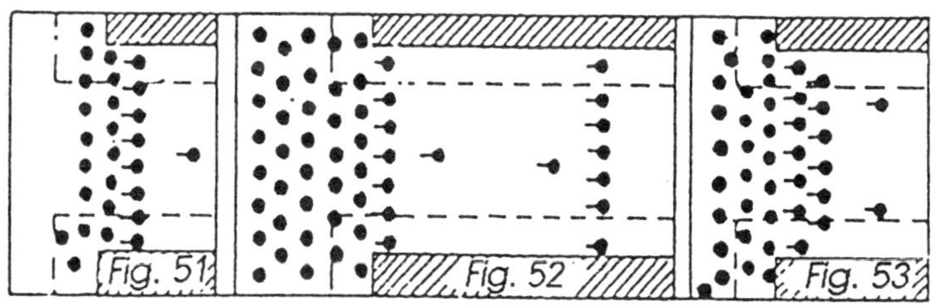
Fig. 51 Fig. 52 Fig. 53

Closing and interdicting much-traveled areas, such as street corners
When the crowd remains stationary and well ordered: patrol behind the crowds (Fig. 56).
When the crowd pushes in one direction:

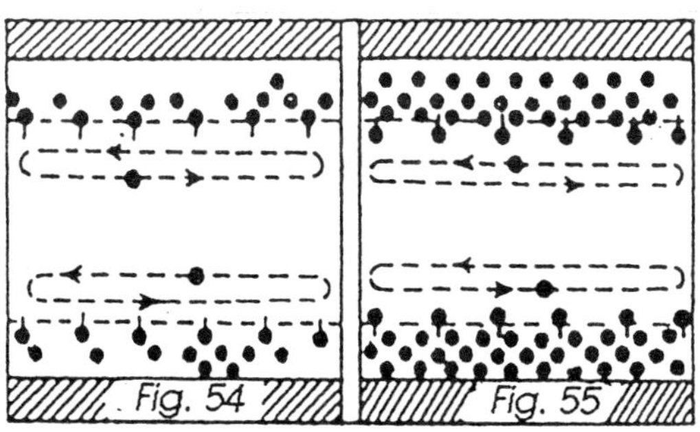
Fig. 54 Fig. 55

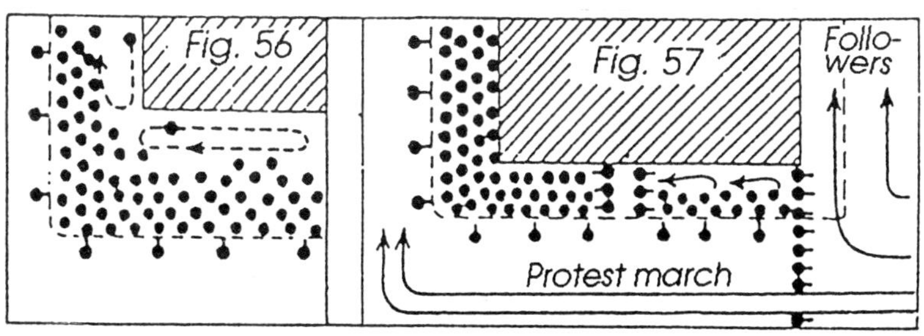

Fig. 56 Fig. 57

form human chains across the sidewalk, to prevent already too-crowded places from becoming more so, as this will endanger the interdiction effort (Fig. 57).

(See also "How to Turn Away Supporters and Followers from an Authorized Demonstration" below.)

The means of closing and interdiction, equipment and armament, as well as the way of holding the carbine (for instance, at the ready), or pistol (holstered or not) and police truncheon (ready or not), depend on the orders in the given situation.

Interdiction by mounted police and the use of police cars

(See Figs. 58 and 59.)

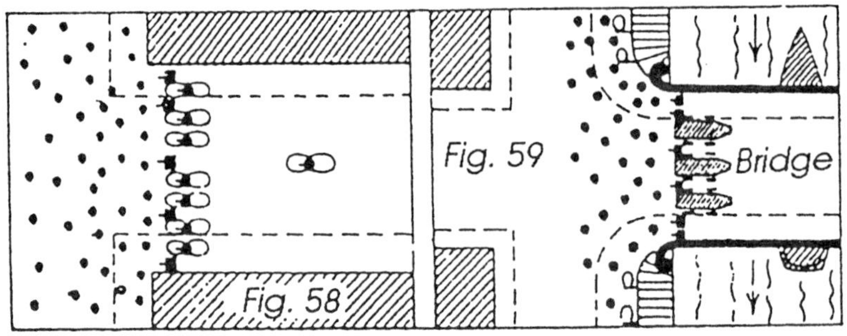

Regulation of civilian vehicular traffic and Streetcars in interdicted areas

Approved traffic driving through an interdicted area may do so only at certain designated locations. The regulation section is preferably commanded by at least a senior police sergeant. Maintain the strictest possible identity control!

Before the start of the interdiction operation, ask for exact orders on where the interdiction is to be. Its extent and what kind of permit is needed for those who have the right to enter or pass through the interdicted area must be decided by the commander.

When necessary, a permit or identity card control post will be set up commanded by a police officer or an experienced official from the Criminal Investigation Police or the Secret State Police (usually behind the roadblock).

Along with the control post, one or more places where traffic is driving through the interdicted area, a place where the closely supervised demonstration march is allowed to pass through, and one or more locations where any crowds may be let through can also be arranged.

When the streetcar network is broken by an interdicted area, streetcars can be allowed through in the following way: the streetcar will stop shortly before reaching the interdiction line. The policemen on the tram car tracks, together with a senior police sergeant detailed for this purpose, then push away the crowds standing on the tracks outside the interdicted zone. When the tracks are free, the streetcar can continue. When the streetcar has passed, the policemen will—with their eyes on the crowd—resume their positions.

Prepare sufficient reserves for any unexpected eventuality.

How to Turn Away Supporters and Followers from an Authorized Demonstration

If the political demonstration or protest march is too big, this method is especially important just before the demonstration reaches its destination, to avoid the public security risk of having large crowds milling about.

Form human chains across the street that will allow the march itself to pass (Fig. 60-a), but will disperse and turn away the followers and hangers-on. When the march has passed, the chain will close the street, and halt the people who want to follow the march (Fig. 60-b). Later, when the march has passed the second chain, then the first chain can be reopened, and so on.

When it seems necessary, additional chains can be formed between the two chains illustrated in Fig. 60.

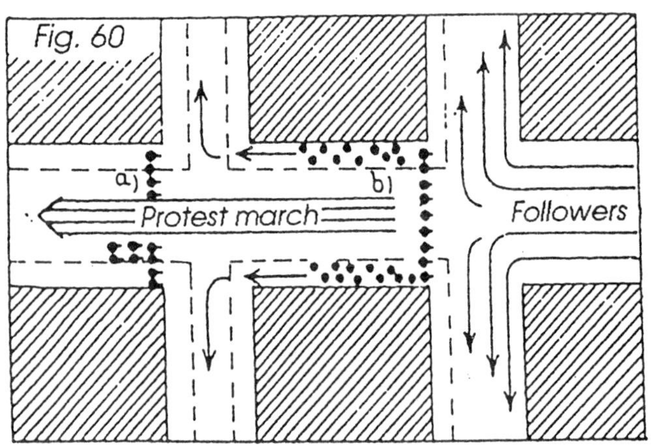

Security during Demonstrations and Protest Marches

When necessary, the march can be protected through interval security posts, accompanying commands, or fast mobile commands with cars (reserves).

Interval security posts

The interval security posts consist of patrols or posts, with police cars stationed along the street of the demonstration and especially at the street crossings.

Accompanying commands

Accompanying commands consist of officers and senior police sergeants, who, on foot or in motor cars, accompany the demonstration. (Mounted bicycle and police are less suitable.) The commands are divided so that some follow at the head of the march, while others follow at the tail, and whenever the conditions so allow, on both sides, or at least along parallel streets. (When so is required, accompanying commands with motor cars can also overtake the march now and then.)

Fast mobile commands (reserves)

Protection is provided through well-prepared, fast mobile commands (reserves) in motor cars. These units will stand in readiness at several nearby positions, next to the planned route of the march, so that they can reach the scene fast in case of need.

How to Direct and Control the Movements of an Authorized Demonstration

The best deployment of available police forces (no stereotyped pattern!)

(1) Sections closing the streets (with minimal rear guard).
(2) Commander and strike team.
(3) Reserves.

1) Sections closing the streets (with minimal rear guard)
(See Fig. 61.)

Task: Interdiction of blocked streets and sensitive areas, where the demonstration or march is not permitted to enter.

Method: Human chain. One or two lines, depending on the situation. The situation also determines use of weapons (carbine, truncheon, and pistol).

2) Commander and strike team
(See Fig. 61.)

Task: To direct and control the march, to break up minor resistance, and so on.

3) Reserves
(See Fig. 61.)

The position of reserves depends on the situation.

Task: Be prepared for any eventuality. Maintain a distant collection place for all arrested civilians and confiscated objects.

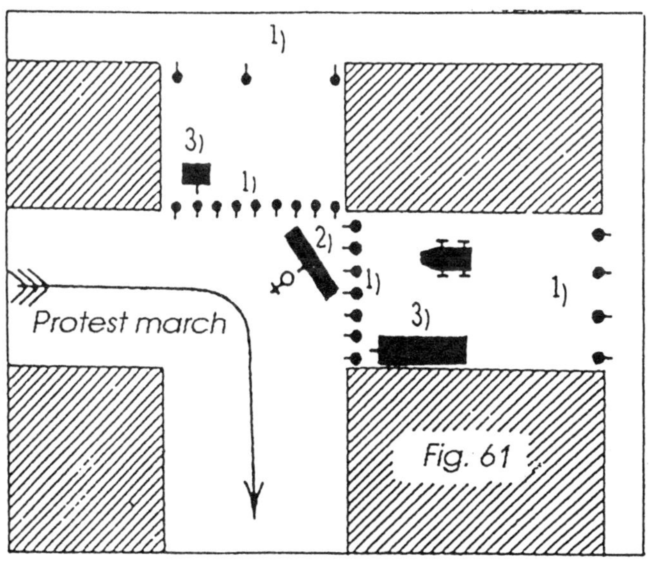

Fig. 61

How to Disperse a Demonstration or Protest March

The best deployment of available police forces (no stereotyped pattern!)

(1) Sections closing the streets (with minimal rear guard)
(2) Dispersal troop
(3) Team for removal of forbidden signs and flags

(4) Evacuation troop
(5) Reserves and collection area for arrested protesters

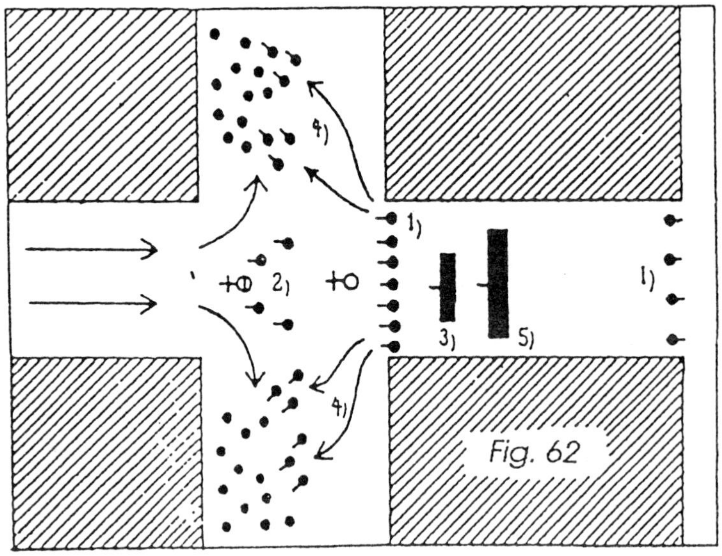

Fig. 62

1) Sections closing the streets (with minimal rear guard)
(See Fig. 62.)
Task: Interdiction of blocked streets and sensitive areas, where the demonstration or march is not permitted to enter. Positioning of a minimal rear guard.
Method: Human chain. One or two lines depending on the situation. The situation also determines use of weapons (carbine, truncheon, and pistol).

2) Dispersal troop
(See Fig. 62.)
Task: To disperse the protest march. Executed by the commander or his assistant commander and a few senior police sergeants.

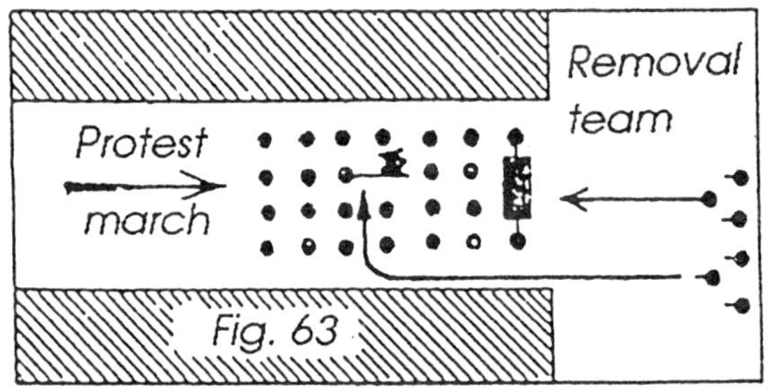

Fig. 63

3) Team for removal of forbidden signs and flags
(See Figs. 62 and 63.)
Task: Confiscation of forbidden signs, flags, etc. If the police force is not strong enough to have a special team for this purpose, the other troops will handle this task at the same time as their regular tasks.

4) Evacuation troop
(See Fig. 62.)
Task: To prevent the formation of larger crowds or groups. Keep the people moving.

5) Reserves and collection area for arrested protesters
(See Fig. 62.)
Task: Be prepared for any eventuality. Maintain a distant collection place for all arrested civilians and confiscated objects.

How to Clear Streets and Public Squares of Crowds

<u>The best deployment of available police forces (no stereotyped pattern!)</u>
(1) Evacuation chain
(2) Patrols for interdiction of side-streets
(3) Rear guard
(4) Reserves
(5) Collection area for arrested civilians
(6) When required: riot force

1) Evacuation chain
(See Figs. 64, 65, and 66.)
Task: Evacuation and clearing the streets and public squares. If one or two lines are used as formation is

ordered by the commander, likewise the position of carbines (in front or at the ready) and the use of truncheons and pistols.

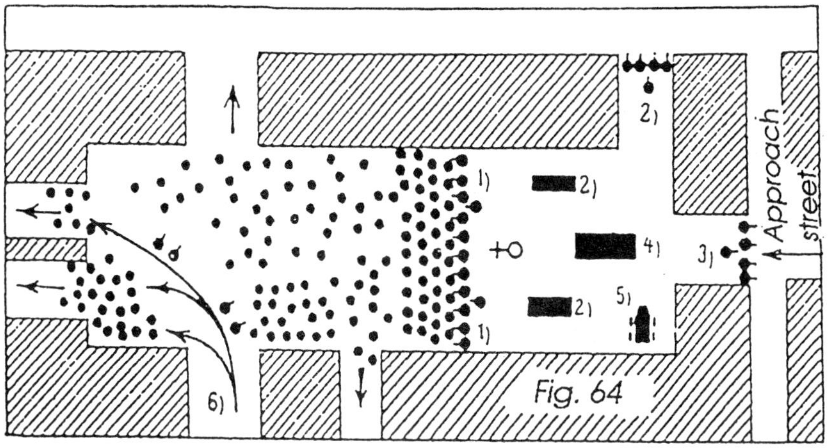

Fig. 64

2) Patrols for interdiction of side-streets

(See Fig. 64.)

Task: Form human chains to interdict disturbances coming from the side streets. The commander will order the shape of the chain, as well as how carbines are to be held, whether pistols are to be holstered or not, and whether truncheons will be ready for use. The best method is to use speedily erected obstacles.

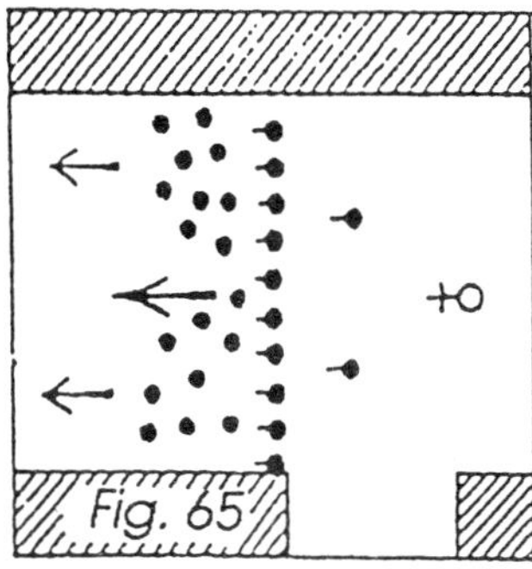
Fig. 65

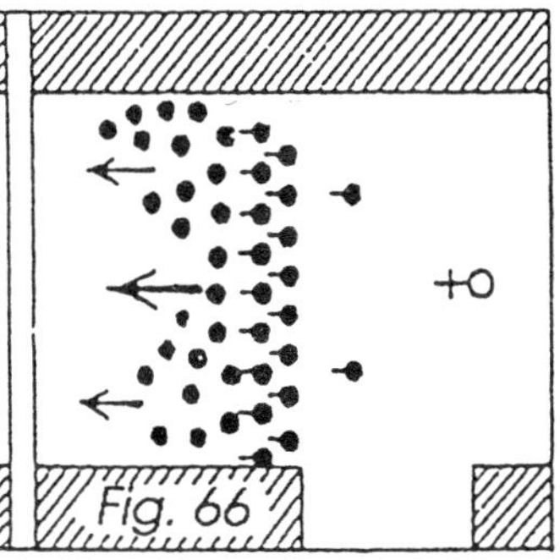
Fig. 66

3) Rear guard

(See Fig. 64.)

Task: Prevention of disturbances at the rear of the evacuation chain. Methods and so on are the same as for human chains (above). The best method is to use speedily erected obstacles.

4) Reserves

(See Fig. 64.)

Task: Be prepared to reinforce the evacuation chain, to break through resisting crowds, etc.

Method: Generally closed march column.

5) Collection area for arrested civilians

(See Figs. 64 and 67.)

The collection area for arrested civilians is usually located where police cars are parked. These are also later used to transport those arrested from the scene. The collection area is commanded and manned by, when possible, senior police officers and officials from the Criminal Investigation Police and the Secret State Police. What possessions to look for and what information to retrieve from the arrested depend on the circumstances, and will be ordered accordingly.

Task: Guarding those arrested. From each person arrested the following must be ascertained and noted: name, reason for arrest, confiscated goods, name of the arresting officer, and witness.

For every confiscated object, a confiscation form with the following information must be filled in:

Confiscated at (place) ______________________________
At (time) ____
Number of objects ________________________________

The police command troop must always bring already prepared arrest forms and confiscation forms in sufficient quantities.

For the organization of collection areas for arrested civilians in times of civil disturbance and unrest, see the next chapter.

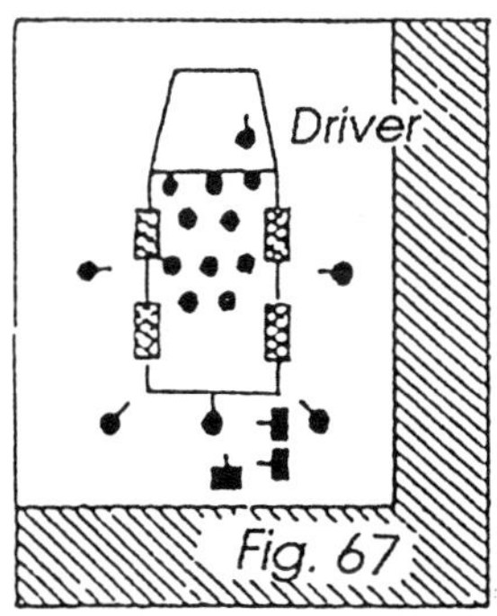

Fig. 67

6) Riot force

(See Fig. 64.)

Task: To split up the crowds and cause the participants to flee by charging into the rear of the crowd. Advance in separate groups from the side streets against the rearward parts of the crowd and force the crowd backward and into other side streets.

This operation can, whenever required, be repeated several times. For this purpose, the use of mounted police is most recommended!

General information

The evacuation is best executed from the narrow side of the public square; in this way we can save our forces, and the crowd has the largest number of escape routes and will therefore disperse faster. A loudspeaker for the use of the commander is required equipment.

How to Transport Prisoners from the Scene of Mass Arrests

The best deployment of available police forces (no stereotyped pattern!)

(1) Police troops

(2) Senior police sergeant in charge of transportation of those arrested from the site of arrest to the local collection area for prisoners

(3) Local collection area for prisoners

(4) Main collection area for prisoners

1) Police troops

(See Fig. 68.)

Those arrested by the police troops will be handed over by the police officers at the site of arrest and sergeants responsible for the arrest to the senior police sergeant in charge of transportation of those arrested—see 2). The sergeant in charge of transportation thereby receives and notes the following information: Cause of arrest, witness (= arresting officer), evidence.

After leaving this information, the arresting officer returns to his section.

2) Senior police sergeant in charge of transportation of those arrested

(See Fig. 68.)

One senior police sergeant is chosen from the police company to be in charge of transportation of those arrested to the local collection area for prisoners.

Task: As soon as an arrest has been made, the sergeant in charge of transportation will take charge of the arrested, make a short note of the information given by the arresting officer, and bring the arrested to the local collection area for prisoners set up by his police unit.

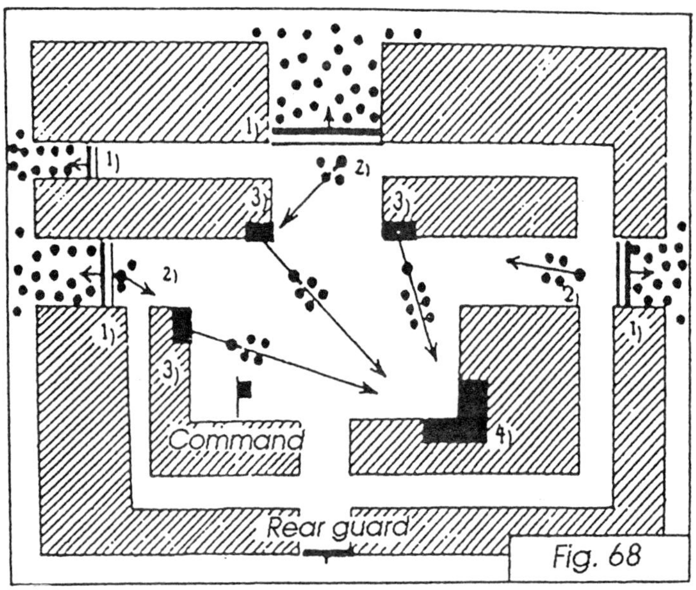

Fig. 68

3) Local collection area for prisoners and confiscated objects

(See Fig. 68.)

The local collection area for prisoners is usually combined with the local collection area for confiscated objects.

Stationed here are the officer in charge of the local collection area for prisoners and confiscated objects (the commander and his command section, officials from the Criminal Investigation

Police and the Secret State Police), a few senior sergeants, and the detail guarding those arrested.

Task: For each arrested, an arrest form (see example below) must be filled in, and the fingerprint of the right index finger of the arrested must be affixed thereon. The arrest form is delivered to the main collection area for prisoners.

When there is time, prepare a special list, in which the name, cause of arrest, witness, and evidence are included, so that the collection area also has a record of those arrested by its police unit.

This section is also responsible for searching the arrested and confiscation of possessions.

When the list is prepared, and there are enough prisoners, these will be brought to the main collection area for prisoners.

For a sample arrest form, see Fig. 69. [*Translator's note*: Below is an English translation.]

Arrest Form

Family and Personal Name: Herz Jakob Georg
Occupation: Locksmith Marital Status: Married
Born on 15. 6. 07 In: Harthof
Town: Rosenheim State: Bavaria
Cause of Arrest: Was today at 14.10 found in No. 17a with a rifle model 98 (No. 250).
Witness: Sgt. Müller I, 4th Police Company, Munich
Evidence: 1 rifle model 98 (No. 250)
Fingerprint of Right Index Finger:
Pfahldorf, 16.3.33
(Place and Date)
Signature of Police Commanding Officer: Stadler

4) Main collection area for prisoners
(See Fig. 68.)

All prisoners taken by all police companies and sections are collected here. Its command consists of one official from the Criminal Investigation Police or the Secret State Police, together with necessary personnel. A state lawyer and an investigation judge are also always appointed. Besides them, the main collection area consists of a senior police sergeant and a guard detail to guard the prisoners.

Task: A thorough police technical investigation of all materials and evidence, through hearing witnesses, maintaining protocol, and investigating every detail. All applicable investigative techniques will be used. In each case, the arrested will be brought to a police headquarters or a prison.

Protection and Security of Authorized Meetings

The best deployment of available police forces (no stereotyped pattern!)
(1) Liaison with state and party officials overseeing the meeting
(2) Section for protecting the meeting
(3) Patrols in the vicinity of the meeting site

1) Liaison with state and party officials overseeing the meeting

Task: Close liaison with the state and party officials overseeing the meeting and the ORPO (Public Order Police) commander.

2) Section for protecting the meeting

Task: Protection of the officials overseeing the meeting and the enforcement of their orders and security

Fig. 69

Festnahmezettel.

Familien= und Vorname: Herz Jakob Georg

Beruf: Schlosser led., verh., verw., gesch.

geboren am 15. 6. 07. in Harthof Gde. Au

Verw.=Bez.: Rosenheim Staat: Bayern

Straftat: wurde heute um 14.10 im Hause Nr. 17a mit einem Gewehr 98 (Nr.250) angetroffen.

Zeuge: Wachtm. Müller I 4. Hdschft. München

Beweismittel: 1 Gewehr 98 (Nr. 250)

Abdruck des rechten Zeigefingers.

Pfahldorf ......, den 16. 3. 33.
(Ort)

Unterschrift: Stadler
Pol.=Hauptwachtmeister.

measures; also protection of the participants in the meeting against dissidents and opponents during and immediately after the meeting. These tasks are most easily accomplished by maintaining guard posts at the meeting site.

"Protection of a closed indoor meeting is most easily achieved by keeping strong police forces close to the meeting area, at best within the building itself and with direct access to the meeting hall."

3) Perimeter patrols in the vicinity of the meeting site

Task: Protection against surprise attacks from the outside. Preventing any crowds from forming when the participants leave after the meeting.

Means of Dispersing an Originally Authorized Meeting

The best deployment of available police forces (no stereotyped pattern!)

(1) Plainclothes officers and undercover officers among the participants of the meeting
(2) Dispersal troops
(3) Evacuation troops
(4) Reserves

(5) Collection area for arrested civilians
(6) Inner security team
(7) Strong forces to prevent the formation of crowds

1) Plainclothes officers and undercover officers among the participants of the meeting
(See Fig. 70.)
Task: Intelligence gathering and timing. Observation of the meeting to call in the security police and regular police forces, when so is required, at the most opportune time. Plainclothes officers will identify the speakers and ringleaders.

2) Dispersal troops
(See Fig. 70.)
Task: To break up the meeting and disperse the participants. These troops are composed of the police commander, one official from the Criminal Investigation Police or the Secret State Police, and a few senior police sergeants.

3) Evacuation troops
(See Fig. 70.)
During the raid the evacuation troops follow closely behind the dispersal troops.
Task: To quickly empty the hall from all participants in the meeting. When an upper gallery exists, at the same time also clear this gallery.

4) Reserves
(See Fig. 70.)
Task: Readiness for any required task.

5) Collection area for arrested civilians
(See Fig. 70.)
Task: As we in this kind of operation (it being an authorized meeting that goes wrong) only expect a few arrests, the best solution is to detain these in a side-room until the end of the operation. Then the prisoners are transported to the police headquarters or prison.

6) Inner security team
(See Fig. 70.)
Task: To prevent the meeting participants from escaping to another story in the building.

7) Strong forces to prevent the formation of crowds
(See Fig. 70.)
Task: To prevent the formation of crowds during or after the operation.

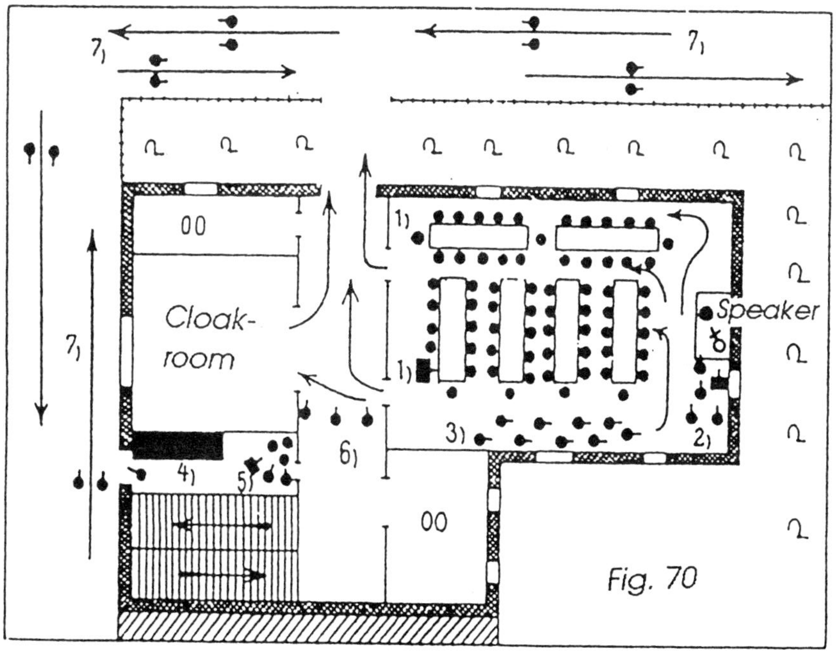

Fig. 70

Raid on a Forbidden or Unauthorized Meeting

<u>The best deployment of available police forces (no stereotyped pattern!)</u>

(1) Reconnaissance and intelligence by plainclothes officers and undercover officers
(2) Outer security teams (outer security ring)
(3) Inner security teams (inner security ring)
(4) Dispersal troops
(5) Evacuation troops
(6) Reserves
(7) Collection area for arrested civilians

1) Reconnaissance and intelligence by plainclothes officers and undercover officers

As long as there is sufficient time, and no immediate danger that this will betray the operation, then a reconnaissance and intelligence effort by plainclothes officers and undercover officers is recommended. If, however, such an effort is impossible, then proceed according to existing maps, house plans, and information from police officials who are familiar with the location, or make up plans just before the operation.

Task: To determine all required intelligence, that is, the exact location and situation in the meeting hall and its neighboring areas (provide a situation sketch map), also the locations of all windows, doors, and so on (provide an observation sketch). Determine the location of and situation in the halls and the cloakrooms of the building. During the meeting, also determine the location of the speaker and the tables where the meeting organizers and ringleaders sit.

2) Outer security teams

(See Fig. 71.)

The outer security teams form an outer security ring around the area of operation.

Task: Prevention of any disturbance from the outside. Deny access to outside supporters. (Also function as rear guard.)

3) Inner security teams

The inner security teams form an inner security ring around the site.

(a) In front of the building:

(See Fig. 71.)

Task: Prevention of any traffic from and to the building. Observation of all doors and windows to prevent important evidence or other objects being disposed of by being thrown from windows or doors. This demands special attention and is an important part of the search procedure during the raid. Also prevent the escape through the windows of any persons inside the building. Along with these tasks, prevent the formation of crowds near the raided building through vigorous patrolling.

(b) Inside the building:

(See Fig. 71.)

Task: Prevent anybody from leaving the building from any except the approved exit. Prevent that people on the other stories disturb the raid on the meeting. Secure that people who were outside the meeting hall at the beginning of the raid are unable to cause disturbances or escape.

4) Dispersal troops

(See Fig. 71.)

Task: Break up the meeting and arrest speakers and ringleaders. These troops are composed of the police commander, one official from the Criminal Investigation Police or the Secret State Police, and a few senior police sergeants. As part of the search procedure, secure important evidence before it is destroyed by the participants in the meeting.

5) Evacuation troops

(See Fig. 71.)

During the raid, the evacuation troops follow closely behind the dispersal troops.

Task: To quickly empty the hall of all participants in the meeting. When an upper gallery exists, at the same time also clear this gallery.

6) Reserves

(See Fig. 71.)

Task: Readiness for any required task.

7) Collection area for arrested civilians

(See Fig. 71.)

Task: As most forbidden or illegal meetings have only few participants, it is best to arrest all people inside the building and transport them to police headquarters or a prison in cars prepared for this purpose.

During raids, every person who cannot identify himself satisfactorily will be arrested and transported to police headquarters or a prison in cars prepared for this purpose.

All those arrested will be searched for illegal belongings.

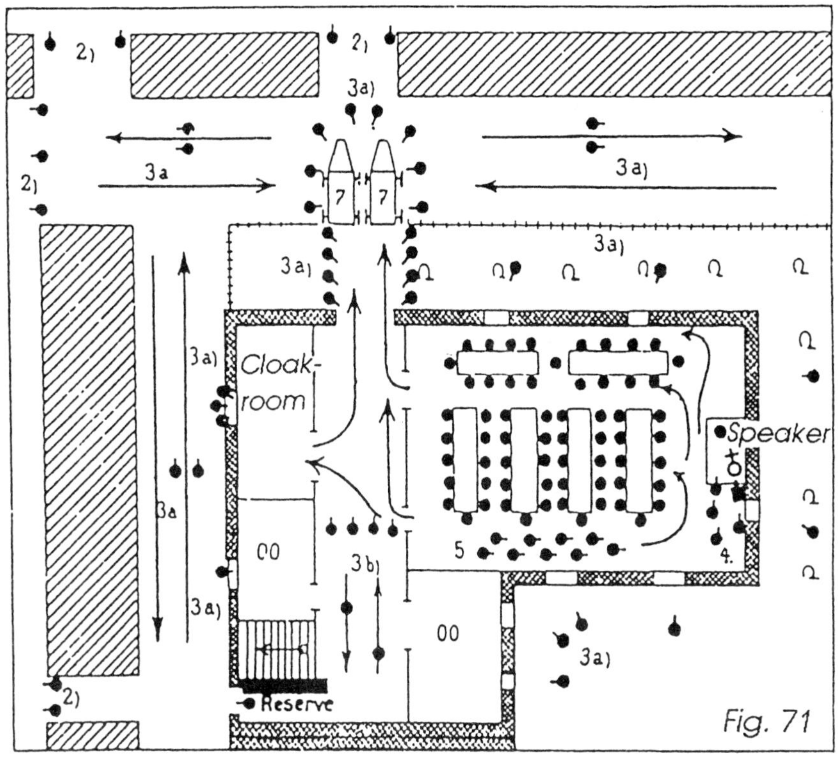

Fig. 71

How to Search a Multistory Building

<u>The best deployment of available police forces (no stereotyped pattern!)</u>

(1) Reconnaissance and intelligence by plainclothes officers
(2) Outer security teams (outer security ring)
(3) Inner security teams (inner security ring)
(4) Search troops
(5) Collection area for arrested civilians and confiscated objects
(6) Reserves

1) Reconnaissance and intelligence by plainclothes officers

Task: To determine the location of and situation in the building and its close neighborhood, also the location of all windows, doors, etc. This information is best presented in a situation sketch map and an observation sketch.

Together with this reconnaissance, the plainclothes officers will also observe the movement of suspicious persons in and out of the building.

Equipment of the plainclothes officer: Only personal clothes, including personal underwear. Do not wear any police uniform items, and especially not any clothes or items with service stamps. Do not bring your service handgun, service bicycle, and so on. No service documents of any kind, including service identification papers, may be carried.

2) Outer security teams

(See Fig. 72.)

The outer security teams form an outer security ring around the area of operations.

Task: Prevention of any disturbance from the outside during the search (also function as rear guard).

3) Inner security teams

The inner security teams form an inner security ring around the site.

(a) In front of the building:

(See Fig. 72.)

Task: Prevention of any traffic from and to the searched building. Observation of all doors and windows, to prevent important evidence or other objects being disposed of by being thrown from windows or doors. This demands special attention and is an important part of the search procedure during the raid. Also prevent persons inside the building from escaping through the windows.

For the deployment of observation posts outside the building, see Fig. 73.

(b) Inside the building:

(See Fig. 72: Corridors, hallways, and stairwells.)

Task: Clear all corridors, hallways, and stairwells. Prevent any movement between the stories. Also prevent any movement from rooms not yet searched into already searched rooms. A number of guard posts must be maintained for these purposes.

4) Search troops

(See Fig. 72: Room C.)

Task: The search troop carries out the actual search procedure. The search troop consists, at best, of officials from the Criminal Investigation Police or the Secret State Police and state judicial auxiliary personnel.

The owner of the building or his representative is to be present during the search. The holder of the apartment lease will be kept under guard during the search and will be present.

Close cooperation between the search troop and the inner security teams is mandatory.

5a) Collection area for arrested civilians and confiscated objects

(See Fig. 72: Room B.)

Task: Guard all arrested persons and objects confiscated as unlawful or as evidence, and be in charge of their transportation. (See "How to Clear Streets and Public Squares from Crowds, 5) Collection area for arrested civilians" above.)

5b) Keeping the inhabitants of the building under guard during the search

(See Fig. 72: Room A.)

Task: Important evidence must be secured before it is destroyed or discarded by the suspects or their associates. In times of civil compliance, all inhabitants of an apartment are told to remain in one room during the search and not to leave that room. One or two senior police sergeants will remain in this room. Under operations during martial law, all civilians found in the building, who are not immediately suspected of a crime and therefore not yet arrested, will be kept under guard by the personnel of the collection area for arrested civilians. The easiest method to accomplish this is to seclude them all in one room, under guard by two policemen.

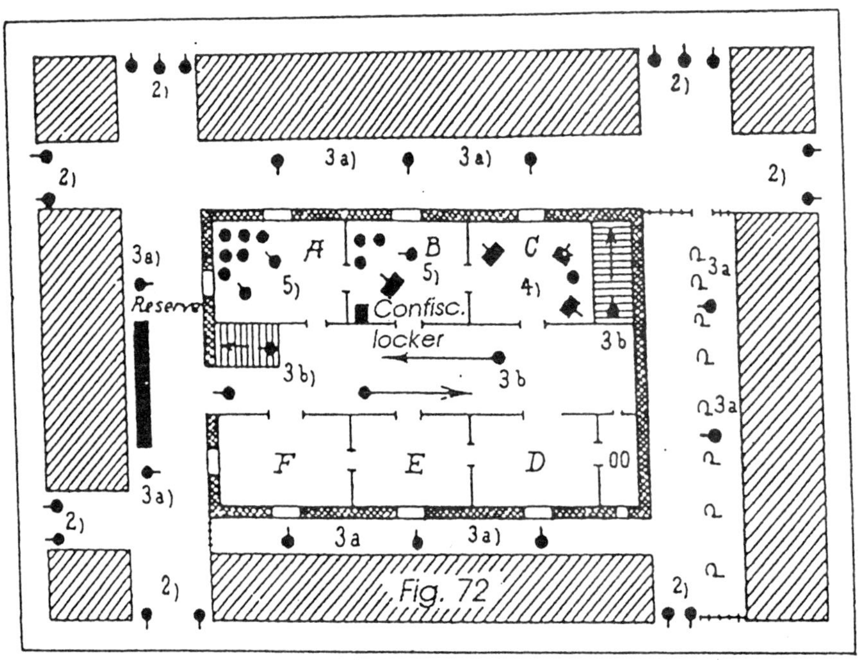

Fig. 72

6) Reserves

(See Fig. 72.)

Task: Readiness for any necessary tasks. The best deployment position is between the inner and the outer security rings.

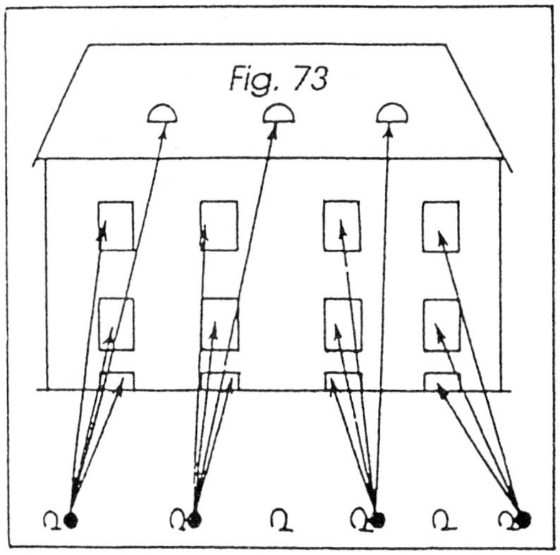

General information

Sufficient equipment for the search, especially electric flashlights and other means for providing necessary light, is absolutely required in any search operation.

Protection and Security of a State or Party Government Building

The best deployment of available police forces (no stereotyped pattern!)

(1) Patrols outside the building
(2) Guard posts immediately in front of the building
(3) Patrols inside the building
(4) Reserves in the guard room

1) Patrols outside the building

Task: By patrolling at irregular intervals provide warning against the appearance of any crowds or suspect persons outside the protected building.

Also reconnaissance of an enemy's possible means of entrance into or attack on the building. Make a special note of the most suitable points for an enemy attack. Under certain circumstances, forward guard posts must be used.

2) Guard posts immediately in front of the building

Task: Prevention of unauthorized entrance into the building. The necessary means to prevent such intrusion will be defined in orders from the commander, depending on the situation. Observation of the immediate surroundings of the building. Identity control of authorized personnel upon entering. A reliable communications procedure with the guard room (see 4) must be secured under all circumstances.

3) Patrols inside the building

Task: To keep the guard posts in contact with each other. Check that doors and windows are closed, and so on.

4) Reserves in the guard room

Task: To take action in case of an attack against the building. Be prepared for any kind of action. Relief of guard posts and patrols.

WARTIME OPERATIONS AND OPERATIONS IN TIMES OF CIVIL UNREST

How to Block Streets

1) Guarded roadblock

Any roadblock fulfills its purpose only when it is also properly guarded. The means and resources allocated to guard a roadblock depend on its importance. Barbed wire is customarily used to interdict a street. In most cases, a guard of 1–2 men is sufficient (Fig. 74). Very important roadblocks are protected by personnel armed with submachine guns or machine guns, and less often also through patrolling police armored cars (Fig. 75); the guard posts stand approximately 20–30 meters behind the roadblock.

If a signboard is used, it should have the text: "Whoever enters will be shot!"

2) Interdiction by the use of submachine guns

In case a street must be interdicted by submachine gun fire, it must first be determined that the fire will not accidentally harm any friendly troops; the most exact geographical limits of the zone of authorized fire must therefore be determined. Most useful is a fire position above the street level (Fig. 76). The commander is responsible for exact and correct information on when fire may be opened.

Movement through City Sections during Civil Disturbance, Rioting, or Enemy Activity

The best deployment of available police forces (no stereotyped pattern!)

(1) Plainclothes officers
Varying distance
(2) Forward police armored car
Varying distance
(3) Advance guard on foot
(4) Flank guard
Distance 20–30 meters
(5) Strike team
(6) Main force
Distance 20–30 meters
(7) Rear guard

Fig. 74

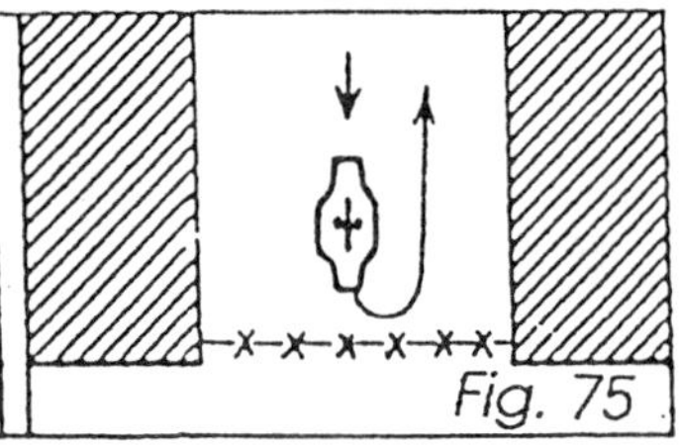
Fig. 75

1) Plainclothes officers

Task: Determine the location of any crowds ready to start a fight, barricades held by the enemy, buildings that appear to be occupied by the enemy, and so on.

2) Forward police armored car
(See Fig. 77.)

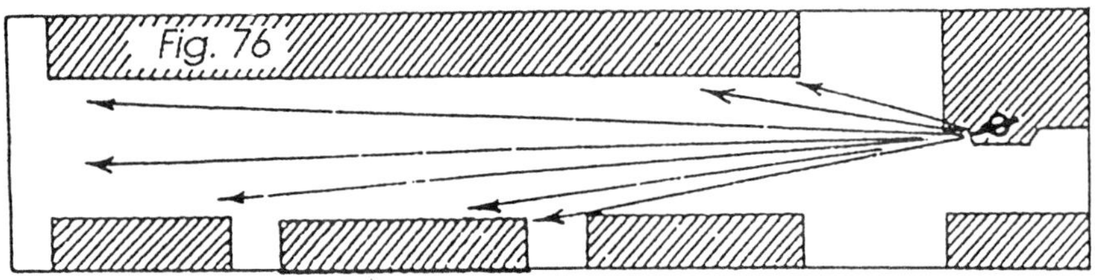
Fig. 76

Transmit any observations toward the rear through signs agreed upon in advance.

Task: Close security—close intelligence gathering and function as an observation platform—protection from enemy surprise attacks.

3) Advance guard on foot
(See Figs. 77 and 78.)

Task: To facilitate the steadiness of the march, to protect the troops of the main force from surprise attacks, and to break through weak enemy resistance. Use submachine guns and machine guns.

Tactics for covering a street with observation and friendly fire during movement: the width of the street decides whether the section marches in column or half-column (as in Fig. 78). Under most circumstances, the carbines will be ready for firing (with safety engaged) and aimed toward the window each man is currently observing.

4) Flank guard
(See Fig. 77.)

Task: To interdict all side streets by establishing flank-guard posts and thus prevent disturbances from the flank. Any approaching enemy or crowd must be immediately reported through one of the flank-guard posts.

When the rear guard passes the last flank-guard post, the flank-guard detail will move on. The commander of the flank-guard detail orders, according to the situation, whether the men of the flank-guard detail will run alongside the marching troops until they reach the flank-guard force proper when the rear guard has passed, or whether they will gather in a troop behind the main force, and only when this troop has reached a sufficient strength will overtake the main force to reach the flank-guard force proper.

5) Strike team
(See Figs. 77 and 79.)
Task: To break any resistance. To deal with barricaded crowds and disperse major crowds.
For an example on how the strike team can break a small resistance group at the flank of the marching force, see Fig. 79.

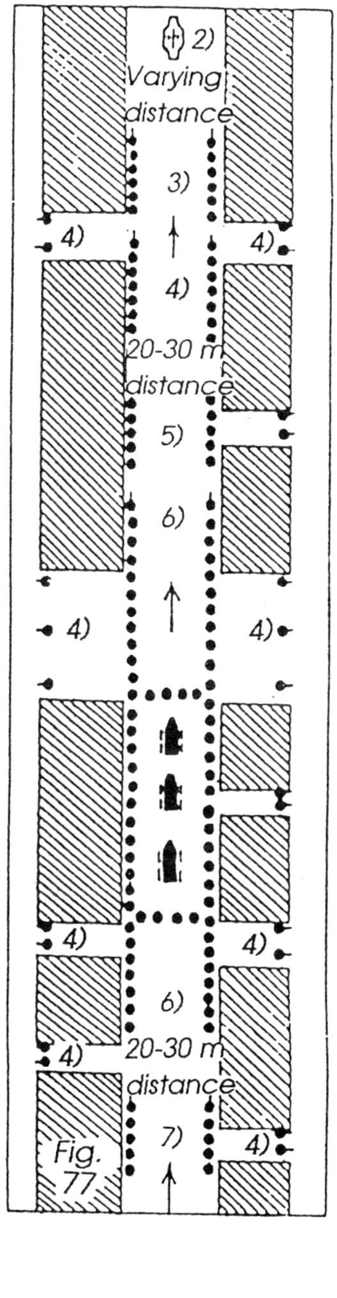

6) Main force
(See Figs. 77 and 80.)
Task: Cover the street with observation and whenever necessary fire during the advance: self-defense through observation of the buildings on the opposite side of the street (the men must be exactly and diligently ordered what each is to observe). In every case, the carbines will be ready to fire, aimed at the window each man is observing.
For deployment and observation duties of the troops, see Fig. 80.
Position any vehicles in the middle of the street. Before and after the vehicles, position police troops formed in lines across the street.

7) Rear guard
(See Figs. 77 and 81.)
Task: Prevention of enemy attack from behind. Use submachine guns and machine guns. Employ proper tactics for covering the street with observation and fire during the advance. For deployment and observation duties, see Fig. 81.

Assault or Raid on an Apartment Block

<u>The best deployment of available police forces (no stereotyped pattern!)</u>
(1) Reconnaissance and intelligence
(2) Outer security teams (outer security ring)
(3) Inner security teams (inner security ring) and patrols
(4) Covering fire
(5) Strike team
(6) Reserves with the collection area for arrested civilians

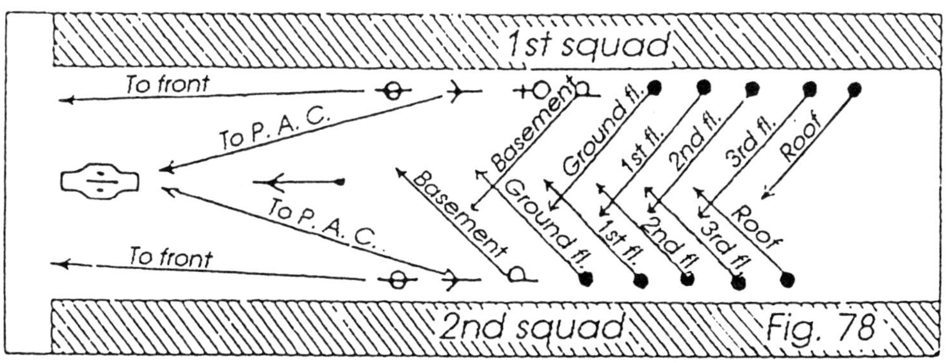

1) Reconnaissance and intelligence

Task: Determine the exact location of and situation in the city block with regard to the neighboring blocks. Report this through a situation sketch map. Report intelligence on the enemy, building layout, etc. Also determine the location of all windows, doors, etc. (in an observation sketch), Determine the strength of the occupying force or crowd; the location of (especially) large numbers of enemies, barricades, and other field fortifications; the enemy's heavy weapons; and how the city block can most easily be approached and entered under covering fire.

2) Outer security teams

(See Fig. 82.)

The outer security teams form the outer security ring around the area of operations.

Task: Prevention of any disturbance from the outside. All roadblocks (barricades or obstacles such as barbed wire) must be guarded and easily protected by covering fire.

3) Inner security teams and patrols

(See Fig. 82.)

The inner security teams form the inner security ring around the building.

Task: To prevent rebels, rioters, or partisans from escaping the building. For this purpose, all approaches to the building on the sides from which no assault will come are closed by covering fire (from submachine guns, whenever possible).

Task of the security patrols: Protect the rear of the strike teams during their approach to the target.

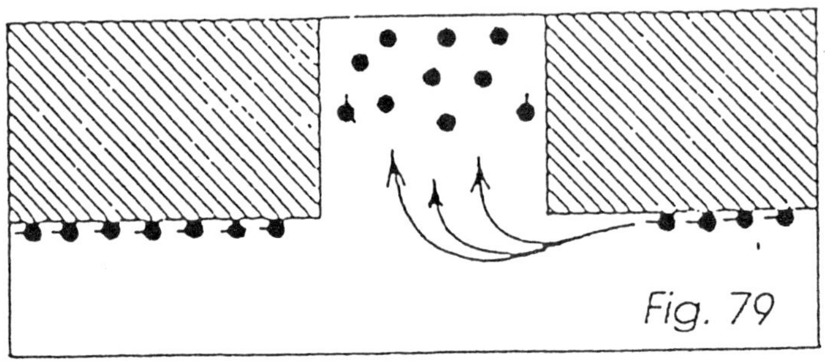

Fig. 79

4) Covering fire

(See Fig. 82.)

Task: The covering-fire teams have the duty to support the approach of the strike team by keeping the enemies down. The covering-fire teams are always put into action whenever the enemy appears or when the strike team advances. It is usually best to position the covering-fire teams on high ground. Maintain reliable communications procedures with the strike team (for instance, with visual light signals).

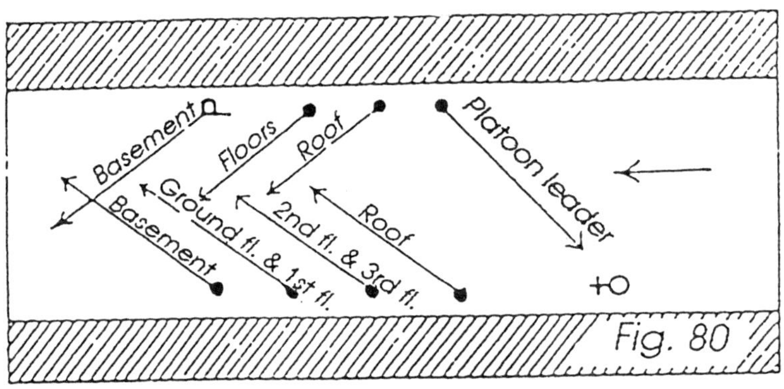

Fig. 80

5) Strike team

(See Figs. 82 and 83.)

Task: The strike team is optimally composed as follows: 1 commander, 1 observer with binoculars and carbine, 1–2 hand grenade throwers with carbines or pistols, 1–2 submachine-gun-armed teams, and a few policemen with carbines; other equipment and

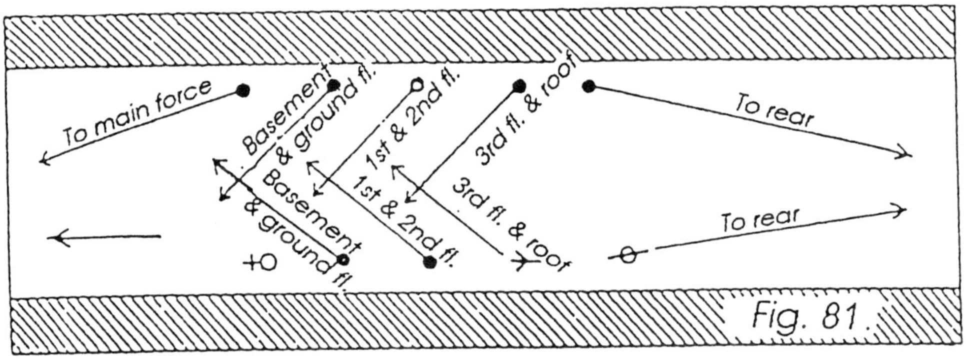

Fig. 81

armament differ according to the situation. Bring what is necessary from the following: "axes, hammers and sledge hammers, heavy-duty tongs, wire cutters, strong rope, ladders, flashlights, searchlights, etc." These tools are used for breaching locked doors, weak walls, and so on. In some circumstances, cooperation with police armored cars is the best solution.

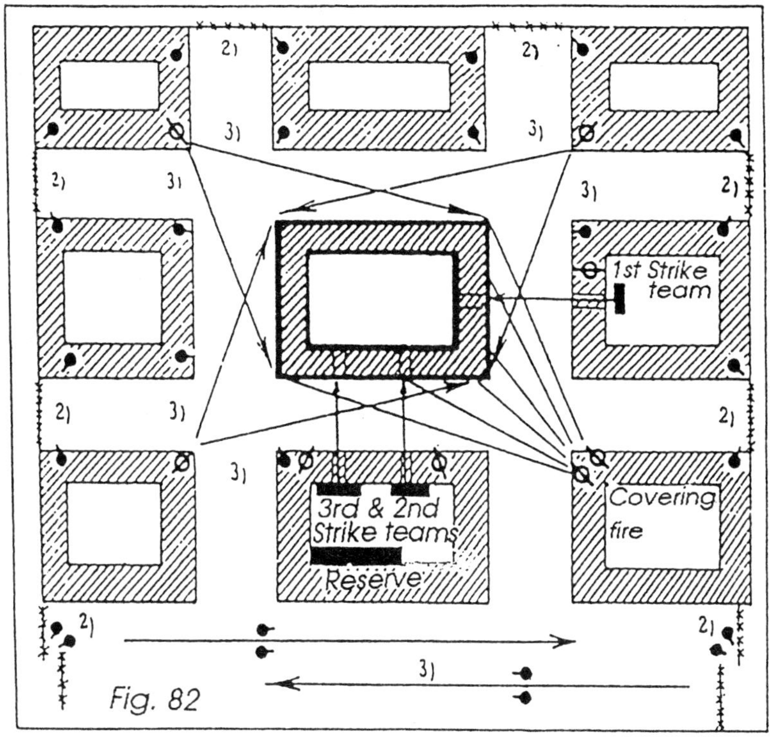

Fig. 82

The grenade throwers will, depending on the situation and the tactical requirements, either throw the grenades or bind them together as linear or concentrated demolition charges and use them as such.

Fig. 83 illustrates the work of a strike team.

Under the protection of the covering fire, as well as the covering fire of the submachine gun team of the strike team, the grenade throwers throw their grenades at the closed door. Whenever possible, however, the door will be demolished with demolition charges, as this is more efficient. After the explosion of the grenades, all members of the troop hurl themselves through the destroyed door into the occupied house. (The covering-fire team of the strike team follows only after a successful entry.)

The strike team must often proceed over roofs, through courtyards and gardens, and, often, also away from the target to safely reach the objective.

When especially difficult obstacles must be crossed, the "clearing troops" of the strike team first take control of the obstacle and clear a path through it. Then the main force of the strike team follows.

6) Reserves

(See Fig. 82.)

Task: The reserves are prepared for any action and provide a guard for the collection area for those arrested.

<u>Tasks after a successful entry</u>

"After the entry into the building, the clearing of the stories begins. To be precise, when the entry takes place at the street level, begin with the lowest story, and when the entry takes place through the roof, begin from the top story." Position interdiction guard posts to secure the operation. The strike team commander must act energetically and quickly reorganize his strike team after the entry.

The search of the building follows according to the section "How to Search a Multistory Building" above.

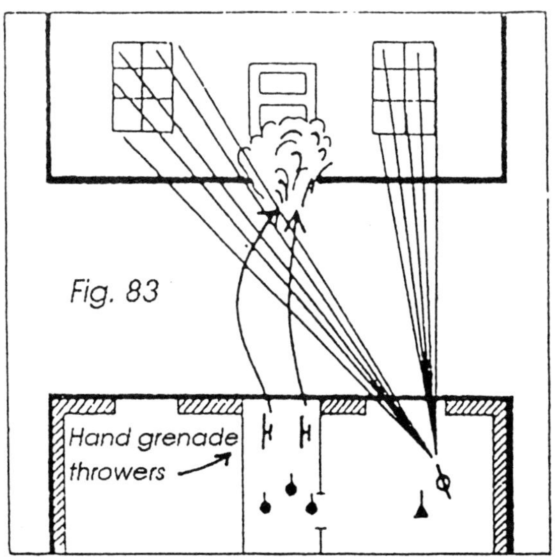

Fig. 83

Pacifying a City Neighborhood

<u>The best deployment of available police forces (no stereotyped pattern!)</u>

(1) Outer security ring

(2) Inner security ring
(3) Building security ring
(4) Search troops
(5) Collection area for arrested civilians and confiscated weapons
(6) Reserves and motor vehicles

1) Outer security ring

(See Fig. 84.)

Task: Interdiction and prevention of any civilian traffic entering or leaving the pacified zone. The security ring consists of an outer perimeter of guard posts within sight and shouting distance of each other. Roadblocks are usually erected. Make certain that all guard posts have appropriate fields of fire! In case of open terrain, maintain security by appropriate concentration of fire. At night, use searchlights.

2) Inner security ring

(See Figs. 84 and 85.)

Task: Separation of the subsectors already searched from those yet to be searched. Prevention of any civilian traffic to and from those subsectors already searched. The security ring is organized as an inner perimeter in the same way as the outer security ring, but denser. Roadblocks are usually erected. Both wings of the inner security ring will connect with the outer security ring.

3) Building security ring

(See Fig. 85.)

Task: Prevention of all civilian traffic to and from those buildings that will be searched, and prevention of all traffic within these buildings. Interdiction and blocking of individual buildings. Observation posts will be positioned.

Compare with the section "How to Search a Multistory Building" above.

Observe the roofs, courtyards, and building fronts with particular care.

Remember that night operations require special equipment.

4) Search troops

Task: Thorough search of all rooms, sheds, and other areas. The search troop optimally consists of 2–4 officials from the Criminal Investigation Police, the Secret State Police, or senior police sergeants. If possible, bring a locksmith (to open locked doors and other locked areas). The formation of many search troops is mandatory when pacifying an entire neighborhood.

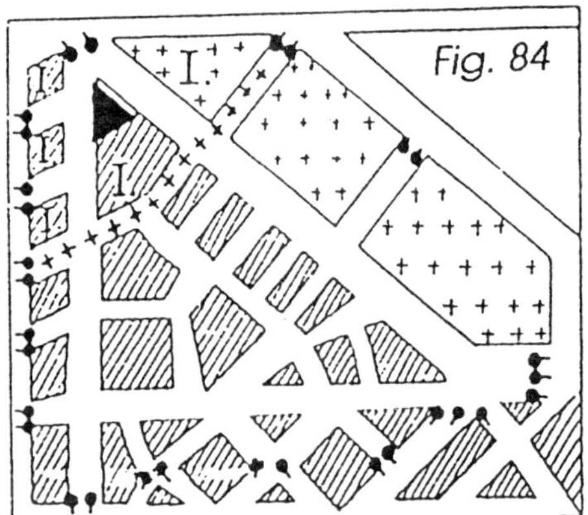

Fig. 84

Outer security ring.
Inner security ring.
Collection area for arrested and reserves.
First sub-sector to be searched.

5) Collection area for arrested civilians and confiscated weapons

(See Figs. 84 and 85.)

Task: To guard arrested civilians and confiscated weapons. All suspicious persons arrested by the police forces are brought here.

This station is commanded by an official from the Criminal Investigation Police or the Secret State Police and a senior police sergeant.

To avoid disturbances of any kind, the best possible location is in a house already searched or its courtyard.

6) Reserves and motor vehicles

(See Fig. 84.)

Task: To save manpower, it is best to incorporate the reserves into the collection area for arrested civilians. Then the reserves can also guard the collection area.

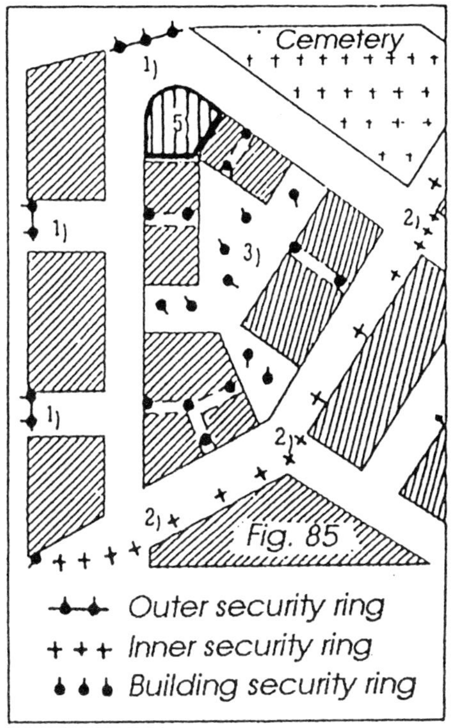

<u>General information</u>

The neighborhood to be pacified is first surrounded by the police. It is then divided into subsectors. Each subsector will be searched in turn.

After searching through subsector I, the outer security ring will assume the previous positions of the inner security ring. The latter advances and takes up positions around subsector II, which is then searched. This deployment method creates a constant border against the subsectors yet to be searched.

After the search is completed in the first subsector, reinforced patrols will resume patrolling the area. The original guard positions (abandoned during the disturbance) are again occupied.

Defense of an Apartment Block

<u>The best deployment of available police forces (no stereotyped pattern!)</u>

(1) Reconnaissance and intelligence patrols
(2) Strong occupation force securing the area in front of the defended city block
(3) Weak security guard posts
(4) Strong reserves (divided into strike teams)
(5) Collection area for arrested civilians

Very important: The best possible and most careful reconnaissance of the defensive conditions of the city block.

1) Reconnaissance and intelligence patrols
(See Fig. 86.)

Besides the reconnaissance of the defensive conditions of the city block mentioned above, you must also aim a reconnaissance and intelligence effort against the enemy.

The task of this reconnaissance and intelligence effort is to determine the strength and armament of the enemy, his plans and chosen direction of attack, the position of his heavy weapons, and so on; also determine the terrain conditions from the enemy's point of view.

2) Strong occupation force securing the area in front of the defended city block
(See Fig. 86.)

The area in front of the defended city block is of special importance in halting the enemy and in making it impossible for him to observe our own defense conditions. The terrain selected for defense positions depends on the situation and the terrain. Make maximum use of submachine guns and machine guns.

The fundamental rule when interdicting this area is to prevent enemy fire from being directed at the defended city block. Make certain that enemy fire is prevented from every position outside the area in front of the defended city block.

Proper use of terrain obstacles or man-made obstacles will save troops.

3) Weak security guard posts (See Fig. 86.)

To keep as many men as possible in the defense against an attack, it follows that the security of the building and the protection of the reserves in it must be made as weak as possible.

Task: The building must be prepared for defense against all directions, with all available means and in the best possible way.

Proper execution: "Windows, balconies, and dormer windows will be fortified with sandbags, rolled carpets or mattresses, cloth, coal boxes, etc. Cloth will be used to hide the defenders against enemy observation. Water must be supplied and prepared against fires. Closed or interdicted approaches, entrances, and obstacles must be under cover of efficient fire, or else such an interdiction is worthless. Tracer ammunition or searchlights must be available and prepared should the attack of rebels, dissidents, or partisans occur at night. Every possibility of a counterattack, also over roofs and through the courtyards of neighboring houses, must be known to all subcommanders."

The following equipment must be kept in readiness and whenever necessary also be used at the right movement: movable barbed-wire entanglements at exits or stairwell entrances; barbed wire for quickly erected obstacles in gardens, etc.; sandbags; flashlights, emergency lights (candles); food and drinking water; reserves of ammunition and hand grenades.

4) Strong reserves (divided into strike teams)

(See Fig. 86.)

To prepare oneself against all possibilities of a resolute attack, strong reserves in the form of strike teams must be formed. The reserves must have good communications with the defense in all directions.

Task: The reserves must be ready for any kind of task. For this purpose, reconnaissance has to be prepared in advance for every conceivable possibility.

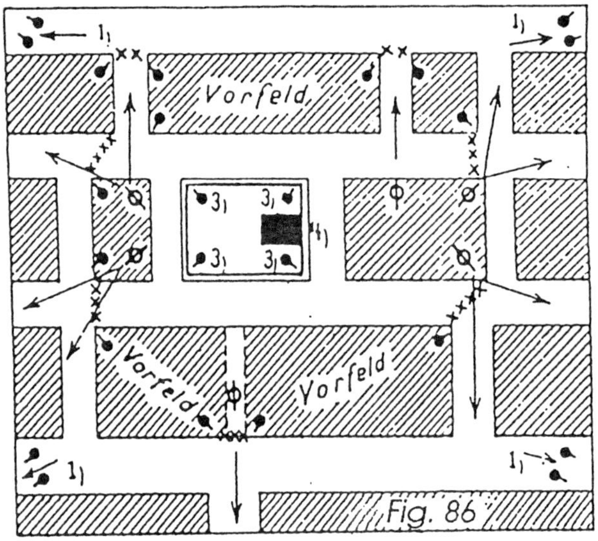

Fig. 86

5) Collection area for arrested civilians

The collection area for arrested civilians is located near the reserves inside the defended city block. All arrested civilians, suspected persons, and enemies are brought here.

Actions taken if the enemy breaks through the defensive position

If the enemy breaks through and enters the building, he must be immediately thrown back by counterattack. The entry must also immediately be reported to the next highest commander.

Organization of Accompanying Security Commands for Transportation of Prisoners, Ammunition, and Provisions

(Procedure when transportation in truck or lorry is impossible)

The personnel are divided into:

(See Fig. 87.)

(1) Wedge-shaped advance and rear guards

(2) Reserves

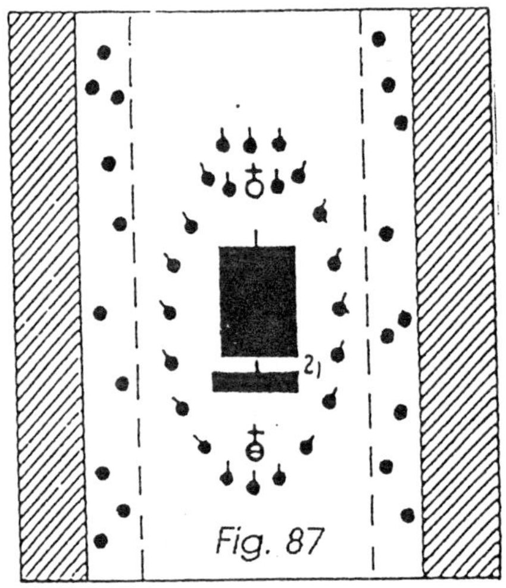
Fig. 87

Task of the security command: Preventing crowds from disturbing the transport. (Hostile crowds commonly attempt to free prisoners or loot ammunition and provisions by rushing the convoy from the sides. Any crowd might be intent on looting a transport.) Minor disturbances can be broken by fast action from the reserves.

It is recommended that you choose a route through side streets with little traffic and that you maintain a forced speed.

APPENDIX 1

MAP SYMBOLS AND ABBREVIATIONS

- Company leader
- Kripo or Gestapo official
- Platoon leader
- Ass. platoon leader
- Senior sergeant
- P — Squad leader
- Orderly
- Policeman w. SMG
- SMG-ammo carrier
- Observer
- Bicycle driver
- MC driver
- Post/Policeman
- Mounted post
- Post with bicycle
- Double post (D. P.)
- Mounted D. P.
- D. P. with bicycles
- Gr.P.1 — Squad post No. 1
- F.W.II. — Field guard detail II
- V.H. — Checkpoint police company
- A.W. — Outer guard detail
- J.W. — Inner guard detail
- W. — Police guard detail

- Q. Kdt. — Town commandant
- Bhf Kdt. — Station commandant
- Police squad
- Police platoon
- Police company
- Sector command
- Mounted section
- Column on foot
- Mounted column
- Column w. bicycles
- Horse-wagon column
- Motor vehicle column
- First aid point
- M.K. — Orderly central
- Radio post
- Visual light signaller
- Searchlight
- Police armored car (moving)
- Cannon
- Heavy machine gun
- Light machine gun

- Fire position (behind weapon symbol)
- For instance, HMG in fire position
- Defensive position
- Defensive point
- Foxholes
- Dummy position
- xxxx — Barbed wire
- x–x–x — Movable barbed-wire entanglement
- YYYY — Log fortifications
- +++++ — Roadblock
- Impassable road
- Destroyed bridge
- Civilian

For other terrain symbols, see the 1:25,000 and 1:100,000 maps.